MICROWAVES

MICROWAVES

(SECOND EDITION)

K C Gupta
Indian Institute of Technology
Kanpur

Publishing Globally

NEW AGE INTERNATIONAL (P) LIMITED, PUBLISHERS
New Delhi • Bangalore • Chennai • Cochin • Guwahati
Hyderabad • Kolkata • Lucknow • Mumbai
Visit us at **www.newagepublishers.com**

Published by New Age International (P) Ltd., Publishers
First Edition: 1983
Second Edition: 2013
Reprint: 2017

GLOBAL OFFICES

• New Delhi **NEW AGE INTERNATIONAL (P) LIMITED, PUBLISHERS**
7/30 A, Daryaganj, New Delhi-110002, (INDIA)
Tel.: (011) 23253771, 23253472, **Telefax:** 23267437, 43551305
E-mail: contactus@newagepublishers.com • Visit us at www.newagepublishers.com

• London **NEW AGE INTERNATIONAL (UK) LTD.**
27 Old Gloucester Street, London, WC1N 3AX, UK
E-mail: info@newacademicscience.co.uk • Visit us at www.newacademicscience.co.uk

• Nairobi **NEW AGE GOLDEN (EAST AFRICA) LTD.**
Ground Floor, Westlands Arcade, Chiromo Road (Next to Naivas Supermarket)
Westlands, Nairobi, KENYA, **Tel.:** 00-254-713848772, 00-254-725700286
E-mail: kenya@newagepublishers.com

BRANCHES

• Bangalore 37/10, 8th Cross (Near Hanuman Temple), Azad Nagar, Chamarajpet, Bangalore- 560 018
Tel.: (080) 26756823, **Telefax:** 26756820, **E-mail: bangalore@newagepublishers.com**

• Chennai 26, Damodaran Street, T. Nagar, Chennai-600 017, **Tel.:** (044) 24353401
Telefax: 24351463, **E-mail: chennai@newagepublishers.com**

• Cochin CC-39/1016, Carrier Station Road, Ernakulam South, Cochin-682 016
Tel.: (0484) 2377303, **Telefax:** 4051304, **E-mail: cochin@newagepublishers.com**

• Guwahati Hemsen Complex, Mohd. Shah Road, Paltan Bazar, Near Starline Hotel
Guwahati-781 008, **Tel.:** (0361) 2513881, **Telefax:** 2543669
E-mail: guwahati@newagepublishers.com

• Hyderabad 105, 1st Floor, Madhiray Kaveri Tower, 3-2-19, Azam Jahi Road, Near Kumar Theater
Nimboliadda Kachiguda, Hyderabad-500 027, **Tel.:** (040) 24652456, **Telefax:** 24652457
E-mail: hyderabad@newagepublishers.com

• Kolkata RDB Chambers (Formerly Lotus Cinema) 106A, 1st Floor, S N Banerjee Road
Kolkata-700 014, **Tel.:** (033) 22273773, **Telefax:** 22275247
E-mail: kolkata@newagepublishers.com

• Lucknow 16-A, Jopling Road, Lucknow-226 001, **Tel.:** (0522) 2209578, 4045297, **Telefax:** 2204098
E-mail: lucknow@newagepublishers.com

• Mumbai 142C, Victor House, Ground Floor, N.M. Joshi Marg, Lower Parel, Mumbai-400 013
Tel.: (022) 24927869, **Telefax:** 24915415, **E-mail: mumbai@newagepublishers.com**

• New Delhi 22, Golden House, Daryaganj, New Delhi-110 002, **Tel.:** (011) 23262368, 23262370
Telefax: 43551305, **E-mail: sales@newagepublishers.com**

ISBN: 978-81-224-3439-2
₹ 299.00
C-16-07-9540
Printed in India at Ajit Printing Press, Delhi.
Typeset at Aakriti Graphics, New Delhi.

NEW AGE INTERNATIONAL (P) LIMITED, PUBLISHERS
7/30 A, Daryaganj, New Delhi-110002
Visit us at **www.newagepublishers.com**
(CIN: U74899DL1966PTC004618)

Dedicated to

USHA

PARUL

SANDEEP

Foreword

It has been a pleasure for me to watch the research and teaching career of Professor K.C. Gupta over the last decade or more. His most recent contribution is his book on *Microwaves.*

This book provides a good coverage of the field for graduate students and research engineers. Apart from classical topics in the field of microwave circuits, measurements, vacuum tube and solid state sources, etc., the coverage extends to non-reciprocal devices, switching elements, and microwave integrated circuits. It also covers some of the emerging industrial applications (as distinct from conventional ones in the field of microwave communications and radar).

The structure of the book is logical and the presentation very lucid. Professor Gupta has brought to this presentation his long experience in the field. In some of the topics such as industrial applications, he has drawn upon his own contributions, not readily available in book form elsewhere.

The written text is supplemented with suitable diagrams. Each chapter has a bibliography and a set of problems at the end. This will help to widen the horizons of knowledge for those users of the book who want to go into greater details, or test their grasp of the subject. It is hoped that the book will be found very useful both by research engineers and graduate students.

December 1978

DR AMARJIT SINGH
Director
C.E.E.R.I.
Pilani 333031

Preface

There have been considerable advances in the field of microwave engineering during last ten years or so. There are new devices available for the generation of microwaves. Amongst other solid state sources, capabilities of GaAs Schottky-barrier-gate field effect transistors are most remarkable. More accurate and simplified methods of microwave measurements especially for network analysis and frequency measurement are now available. Yttrium iorn garnet resonators now play a significant role in tunable filters and oscillators. PIN diode controlled components are frequently used for microwave switching, phase-shifting, attenuating, modulating and limiting. Subnanosecond logic and pulse circuits using Gunn devices and/or GaAs MESFETs are on the threshold of finding system applications. Hybrid integrated circuits are commonly used in microwave systems. Lumped inductors, capacitors and resistors are becoming popular at microwave frequencies and promise to eliminate the barrier between circuit design philosophy for 'microwaves' and 'lower frequencies'. Also, microwaves are now finding interesting applications in several non-communication areas like industrial control and measurement and dielectric heating at microwave frequencies. All these advances have brought definite changes in the courses of microwaves at undergraduate and postgraduate levels. A basic understanding of newer aspects of microwave engineering is necessary for students interested in this area. The present text is the result of the attempts to incorporate these new trends in the first course on 'Microwaves'.

This book has evolved from the lecture notes prepared for teaching a course on 'Microwaves' to final year undergraduate and postgraduate students of electrical engineering at the Indian Institute of Technology, Kanpur, during the period 1969 to 1975. Absence of an equivalent textbook for the first course on microwaves has been the main motivation for the preparation of this text. Bound notes emerged in November 1974 which were revised in July 1975. Considerable revision and updating of the notes was then done when the author was on an extended leave of absence at University of Waterloo, Waterloo (Canada), Ecole Polytechnic Federale de Lausanne (Switzerland) and Technical University of Denmark (at Lyngby) during 1975 to 1977. Comments by several colleagues and by the reviewers of National Book Trust, India, have been helpful in shaping the present version.

At I.I.T., Kanpur, the course on Microwaves follows a prerequisite introductory course on 'Electromagnetic Theory' in pre-final year. Maxwell's equations, transmission lines, waveguides and resonators are discussed in the Electromagnetic Theory course. However, a concise description of these topics has been included in Chapter Two. This chapter provides a necessary base for several topics discussed later and may be considered optional in lecture (or study) schedule. Usually a course on

Microwaves has a laboratory associated with it. This allows the students to pick up measurements techniques during the laboratory work. A brief discussion on microwave measurement is, however, included in Chapter Six which describes basic techniques and principles of important microwave instruments. The problems at the end of various chapters are designed to reinforce the key aspects and are strongly recommended to be worked out by students. 'Select Bibliography' at the end of each chapter lists the important sources where a more detailed treatment of the material is available. Several topics related to microwaves, such as antennas, propagation, radar systems, communication systems and low-noise amplifiers which are discussed in other elective course at final year level are not included in this text.

Several colleagues, students and friends have contributed to the preparation of this book by providing reviews, comments and criticisms. These contributions and especially those of Dr I.J. Bahl, Dr Ramesh Garg and Mr Rakesh Chadha are gratefully acknowledged. The author is thankful to Professor K.R. Sarma for his interest and encouragements for this work. Facilities extended by the Department of Electrical Engineering and the Advanced Centre for Electronic Systems at I.I.T., Kanpur are acknowledged. The typing of manuscript has been handled efficiently by Mr C.M. Abraham and the artwork has been done by Mr R.K. Bajpai. The preparation of this manuscript has been supported by Educational Development Centre, I.I.T., Kanpur, through their projects No. EDC/N-EE/4 (1974-75) and EDC/KCG-3/EE (1977-78). This support is thankfully acknowledged. Finally, I would like to thank my wife Usha for her patience and understanding throughout the various stages of evolution of this manuscript.

K C GUPTA

Kanpur
November 1978

Contents

CHAPTER 1

Introduction

1.1 WHAT ARE MICROWAVES?

'Microwaves' is a descriptive term used to identify electromagnetic waves in the frequency spectrum ranging approximately from 1 GigaHertz (10^9 Hertz) to 30 GigaHertz. This corresponds to wavelengths from 30 cm to 1 cm. Sometimes higher frequencies (extending up to 600 GHz) are also called 'microwaves'. These waves present several interesting and unusual features not found in other portions of the electromagnetic frequency spectrum. These features make 'microwaves' uniquely suitable for several useful applications.

1.2 CHARACTERISTIC FEATURES OF MICROWAVES

The main characteristic features of microwaves originate from the small size of wavelengths (1 cm to 30 cm) in relation to the sizes of components or devices commonly used. Since the wavelengths are small, the phase varies rapidly with distance; consequently the techniques of circuit analysis and design, of measurements and of power generation, and amplification at these frequencies are distinct from those at lower frequencies.

For dealing with these small wavelengths, methods of circuit representation and analysis need to be modified. The phase difference caused by the interconnection between various components or various parts of a single component is not negligible. Consequently, analyses based on Kirchhoff's laws and voltage-current concepts are not adequate to describe the circuit behaviour at microwave frequencies. It is necessary to analyse the circuit or the component in terms of electric and magnetic fields associated with it. For this reason microwave engineering is also known as electromagnetic engineering or applied electromagnetics. A background of electromagnetic theory is a pre-requisite for understanding microwaves.

Not only analytical techniques, the methods of measurement also become specialized at microwave frequencies. Measurements are carried out in terms of field amplitudes, phase differences and powers carried by the waves. A very commonly used method of microwave measurements is based on the study of a standing wave pattern formed along the line because of the interference of incident and reflected waves. Ratio of the amplitudes and phase relationship between incident and reflected waves tell us about impedance characteristics of the components causing the reflection. Several other special techniques have been developed for use at microwave frequencies.

Microwave circuit components also have a different look. Use of lumped elements at microwave frequencies becomes difficult because of small wavelengths involved. For realizing the lumped behaviour of a capacitor, an inductor or a resistor, the component size must be much smaller than the wavelength. Because of this reason microwave systems employ distributed circuit elements very often. These elements are made up of small sections of transmission lines and waveguides. For example, a quarter wavelength section of a transmission line is used as an impedance transformer. A half wavelength section, on the other hand, constitutes a resonant circuit to be used in place of an L-C resonator. Use of miniature lumped elements at microwave frequencies has been made possible by the advancement in micro-electronics technology during the last ten years. These techniques allow us to fabricate inductors, capacitors and resistors that are about a millimeter or two in size; much smaller than small wavelengths at microwave frequencies. Another aspect unique to microwave circuits is the possibility of radiation from discontinuities in distributed circuits. This necessitates a more careful and accurate circuit design.

The challenge of generation of microwaves has resulted in a variety of devices—both in vacuum tube and in semiconductor device areas. When an attempt is made to use a lower frequency source (triode or transistor) at microwave frequencies, the operation is limited by the fact that transit time of the carriers through the device (i.e. electrons in triodes and electrons or holes in transistors) becomes comparable to the time period of the wave. This problem has been solved by technological innovations (in case of transistors) and by totally novel ideas (as in case of klystrons, magnetrons, transferred electron devices and avalanche diodes).

1.3 APPLICATIONS OF MICROWAVES

Study and research in microwaves has not only been an interesting and challenging academic endeavour, it has led to several useful applications in communications, in radar, in physical research, in medicine and in industrial measurements and also for heating and drying of agricultural and food products.

A significant advantage associated with the use of microwaves for communications is their large bandwidth. A ten per cent bandwidth at 3 GHz implies availability of 300 MHz spectrum. This means all the radio, television and other communications that are transmitted in the frequency spectrum from DC to 300 MHz can be accommodated in a ten per cent bandwidth around 3 GHz (say from 2850 to 3150 MHz).* Since the lower frequency part of the radio

---

* Approximately fifteen hundred FM radio channels or fifty television channels could be broadcast in this frequency range.

spectrum is getting crowded, there is a trend to use more and more of microwave region (and beyond) for various different services.

Presently, microwave communications are widely used for telephone networks, in broadcast and television systems and in several other communication applications by services, railways, etc.

Short wavelengths also simplify the design and installation of high directivity antennae. Antenna directivity depends on the ratio of antenna aperture to the wavelength of the signal to be transmitted. At 10 GHz, a pencil beam with 1° beamwidth can be obtained by using a 6.9 feet diameter antenna. At 10 MHz, this will require an antenna diameter of 6,900 feet. This becomes impractical, specially if it is desired to rotate the antenna so that the beam can look in various directions.

Small antenna size and the property of reflection of microwaves from metallic surfaces make it practical to operate radar systems at these frequencies. Radar is an electronic method of detecting the presence of aircraft (or other objects) at ranges and in circumstances where other means of detection are not possible. Operation of the radar is based on the measurement of the time it takes for a pulse transmitted from an antenna to get reflected by the object to be detected and to return at the antenna and the receiver. Also, in many radars, a shift in the frequency of reflected signal caused by the Doppler effect can be recorded. Velocity of the target can be calculated from this measurement. The reflection from the object to be measured is significant only when the wavelength is much smaller than the size of the object. For this reason, the radar could not become practicable at lower frequency and had to wait for the development of microwave technology during the Second World War period. Today, radars constitute about 70 per cent of microwave equipment. There is a whole variety of radars: early-warning radar, missile-tracking radar, missile-guidance radar, fire-control radar, weather-detection radar, air-traffic-control radar and even radars to detect and control the speed of automobiles.

There are other advantages associated with the small wevelengths at microwave frequencies. Unlike lower radio frequencies, these waves are not reflected and practically not absorbed by the ionosphere. This has led radio astronomers to use these frequencies to study electromagnetic radiations originating from stars and other astronomical objects. Also, this property makes microwaves suitable for space communication and satellite communication.

Microwave techniques are now being introduced in extremely fast computer operations. Pulses with very small widths are used in high speed logic circuits. When the pulse width lies in the sub-nanosecond range, the major portion of the pulse spectrum is in the microwave region. Thus, microwave techniques are useful in transmission and handling of these pulses. These sub-nanosecond pulses are also useful in several special purpose radars for docking the pre-collision sensing, etc.

Microwaves exhibit another interesting feature. Molecular, atomic, and nuclear systems display various resonance phenomena when placed in periodic electromagnetic fields.* Several of these resonance absorption lines lie in the microwave frequency range. The resonance absorption is due to rotational transitions in the molecules and the absorption spectra provide information on the molecular structure and intramolecular energies. Thus, microwaves become a very powerful experimental tool for the study of some of the basic properties of materials. Besides scientific research, absorption of microwaves by molecular resonances is well suited for various industrial measurements. It can be used to measure the concentration of different gases, e.g. in exhaust chimneys in order to control the emission of pollutants, or in chemical processes in order to record continuously the concentration of gases evolved in the process.

The study of microwave resonances in molecules has led to several useful devices. The most significant ones are the non-reciprocal devices employing ferrites and soild state microwave amplifiers and oscillators called masers. The magnetic properties of microwave ferrites are due to the electron spins in solids. The coupling between spins is such as to divide the magnetic atoms into groups having oppositely oriented magnetic dipoles. When placed in an external static magnetic field these materials exhibit non-reciprocal behaviour at microwave frequencies. The maser is a microwave amplifier or oscillator which employs, as its working substance, a paramagnetic material having a suitable set of electron-spin energy levels separated by energy intervals that correspond to the frequencies in the microwave range. Stimulated transition from a higher energy state to a lower state results in radiation at microwave frequencies. Masers have lower noise figure than any other type of amplifiers known and are used in the communication systems where extremely low noise characteristics are desired.

Just like any other form of energy, microwave energy can also be used for heating. Thermal effects produced by microwaves have a variety of industrial applications. Microwave ovens for cooking follow the principle of dielectric heating. Cooking is done very quickly and uniformly by microwaves since the food is cooked by the waves on the inside at the same time as on the outside. Like transfer of heat by conduction, convection and radiation, microwave heating can be considered as another mode of heat transfer. In this mode, heat is produced directly at the locations of the dielectric losses. Water has higher dielectric loss than the other ingredients in food products. Thus, water pockets get heated first which is exactly where heat is required for cooking purposes. Microwave diathermy machines produce heat inside the muscles without heating the tissues and skin outside. Also, microwave drying machines are used in printing, textile, and paper industries.

---

* The resonant absorption is accompanied by the usual non-resonant absorption of electromagnetic waves in various materials.

1.4 OUTLINE OF THE BOOK

The material in the rest of this book is divided into eleven chapters.

Chapter Two describes the characteristics of transmission structures and resonators used at microwave frequencies. This description is intended to provide concise revision of what is normally available in detail in text-books on Electromagnetics. Basic transmission line equations are obtained by writing equivalent circuit for a small section of the line. Concepts of reflection coefficient, standing waves and input impedance are derived from these transmission line equations. Solution of transmission line problems can be carried out conveniently by using a graphical aid called Smith chart. Transmission line sections are also used as circuit elements at microwave frequencies. Rectangular and circular waveguides are two commonly used transmission structures for microwaves. Their characteristics are obtained by solving the wave equation. This leads to two possible modes of propagation called transverse electric (TE) and transverse magnetic (TM) modes. Attenuation coefficient and excitation methods are discussed for the case of rectangular waveguides. Small sections of transmission lines and waveguides can be used as reactances and resonators. Rectangular and circular waveguide resonators are also analysed by solving the wave equation with appropriate boundary conditions. Derivation of quality factor for rectangular waveguide resonators is included.

The next two chapters deal with microwave sources. Generation of microwaves by vacuum tubes forms the subject matter of Chapter Three. Use of conventional tubes (like triodes) at microwave frequencies is limited because of stray reactances and transit time effects. Working of klystron amplifiers used at microwave frequencies has been presented in terms of velocity and density modulations of an electron beam. Multicavity klystrons used for amplifiers, and single cavity reflex klystrons used as oscillators are discussed. Another microwave tube, called Magnetron, is based on the dynamics of electrons in crossed electric and magnetic fields. A qualitative picture of oscillation mechanism in magnetrons has also been presented. Travelling wave tubes use cumulative interaction between space charge waves on electron beams and electromagnetic waves on slow wave structures. An analysis of coupling between the beam and the slow wave circuit is described. Backward wave oscillator is a special version of travelling wave tube and is included in this chapter.

Chapter Four contains a description of microwave sources using semi-conductor devices. The discussion starts with a presentation of the relative performance of various active semiconductor devices operating at microwave frequencies. Various features that distinguish microwave bipolar transistors from the transistors at lower frequencies are pointed out. Microwave field effect transistors developed during the past few years use GaAs technology and a Schottky barrier junction for the gate contact. Other devices discussed include transferred electron devices (Gunn diodes) and avalanche diodes. Circuit aspects of the oscillators using these devices are also discussed.

Chapter Five presents a brief description of the basic concepts useful for representation of microwave networks. Kirchhoff's laws, that are the basis for the analysis of electrical networks, are shown to be consistent with Maxwell's equations. Voltages and currents cannot be defined uniquely for TE and TM modes in waveguides. However, waveguide impedance is defined in terms of the power flow and a voltage which is defined in a somewhat ad hoc manner. Scattering matrix formulation is widely used for microwave networks. Characteristics of scattering matrices are discussed and several cxamples of typical networks are presented.

Basic measurement techniques useful at microwave frequencies form the theme of Chapter Six. Detection of microwaves by point contact and Schottky barrier diodes and the measurement of microwave power by bolometers are described. Techniques of impedance measurement discussed include slotted line method and the use of network analyser. Measurement of scattering parameters for multi-port networks is carried out by using directional couplers and a ratio meter (when only magnitude of coefficients is needed) or a network analyser (when the phase information is also required). Various methods have been described for frequency measurement. These include transfer oscillators and direct reading microwave frequency counters.

Chapter Seven deals with three most commonly used passive components in microwave circuits, namely, impedance transformers, filters and directional couplers. Impedance transformation (or impedance matching) can be obtained using stubs or quarter-wave sections. Both of these methods are discussed and multi-sectional quarter wave transformers have been given a detailed treatment. The analysis of multi-sectional transformers is based on small reflections approximation. These transformers can be designed for maximally flat or Chebyshev response. Filter design procedures discussed include derivation of filters from multi-sectional quarter wave transformers and filter design based on low frequency prototype design. Discussions on directional couplers include: multi-hole directional couplers, scattering matrix for directional couplers, coupled line directional couplers and matched hybrid-tee.

Chapter Eight deals with non-reciprocal devices using ferrite material. Permeability tensor for a magnetized ferrite medium is derived from Maxwell's equations. Wave propagation in ferrite media is discussed in terms of circularly polarized waves. The phenomenon of Faraday rotation is used for describing the operation of isolators, four-port circulators, switches and modulators. A phenomenological description of a three-port circulator has also been included. Resonance phenomenon in ferrite finds applications in isolators. Also, resonance in single crystal garnets (YIG, etc.) is used to obtain tunable resonators.

Chapter Nine includes discussions on PIN diode control components and ultrafast logic circuits using sub-nanosecond pulses. These pulses have a major part of their frequency spectrum in microwave range and thus microwave circuit techniques are applicable. Operation of PIN diodes is different from that of *p-n* junction diodes. Also included are the applications of PIN diodes in various types of circuits such as switches, phase shifters, attenuators, modulators and limiters. Transferred electron devices and GaAs field effect transistors are useful for sub-nanosecond logic circuits. A few examples of these circuits, both combinatorial and sequential, are also given.

Chapter Ten on 'Microwave Integrated Circuits' includes the descriptions of the planar transmission lines and the technology of hybrid MICs. Transmission lines suitable for use in MICs include striplines, microstrip lines, slot lines and coplanar lines. Hybrid MIC technology uses photo-lithography and thin film techniques. Advantages of MIC approach and various difficulties encountered with MICs are also discussed.

Use of lumped elements at microwave frequencies is the topic of Chapter Eleven. Small size inductors and capacitors can be designed for use at microwave frequencies. Design and fabrication of these elements for use in hybrid MICs is described. Special measurement techniques have been evolved for characterization of these lumped elements. Also discussed are the use of these elements in circuits and the resulting advantages.

'Industrial applications' of microwave are discussed in Chapter Twelve. Topics described therein include: industrial control and measurements, doppler motion sensors and microwave heating. There are brief descriptions of thickness measurement of metal sheets in rolling machines, continuous measurement of wire diameters and monitoring and measurement of moisture content. Doppler motion sensors are based on the principle of doppler radars and are discussed briefly. Microwave energy is also used for heating of dielectric materials. Basic principles involved and design of applicators have also been explained.

CHAPTER 2

Transmission Structures and Resonators

In addition to two-wire and coaxial transmission lines commonly employed at lower frequencies, a type of transmission structures called waveguides is widely used at microwave frequencies. Waveguides are single conductor, rectangular or circular tubes capable of guiding high frequency electromagnetic waves through them. Apart from their use for transmission of microwaves, short sections of transmission lines and waveguides are also used as circuit elements (equivalent to inductors and capacitors) at these frequencies. Resonant circuits for use at microwave frequencies are also derived from these transmission structures. This chapter is a brief review of transmission lines, waveguides and the resonators derived therefrom.

2.1 TRANSMISSION LINES

2.1.1 Analysis

Transmission lines consist of two or more conductors used for carrying electromagnetic waves from one point to another. They are available in various configurations. The most common forms are : coaxial and two-wire lines. These are shown in Fig. 2.1 (a). Because of its shielded nature, the coaxial line is preferred for use at microwave frequencies. These transmission lines may be analysed by representing a small section of the line by an equivalent lumped element circuit shown in Fig. 2.1(b).

Values of inductance L and capacitance C per unit length are calculated from magnetostatic and electrostatic field considerations. For good transmission lines the series resistance per unit length R and the shunt conductance per unit length G are negligible. For coaxial line configuration

Fig. 2. 1(a), parameters L and C are given by the following relations

$$L \text{ (henry/m)} = \frac{\mu}{2\pi} \text{ In } \frac{b}{a} \qquad \text{...(2.1a)}$$

and

$$C \text{ (farads/m)} = 2\pi\varepsilon \text{ In } \frac{b}{a} \qquad \text{...(2.1b)}$$

where μ and ε represent the permeability and permittivity, respectively of the dielectric medium between the two conductors. Symbols b and a denote radii of outer and inner conductors, respectively. These values of L and C are used for finding characteristic impedance and phase velocity of coaxial line by using the formulae derived later in this chapter.

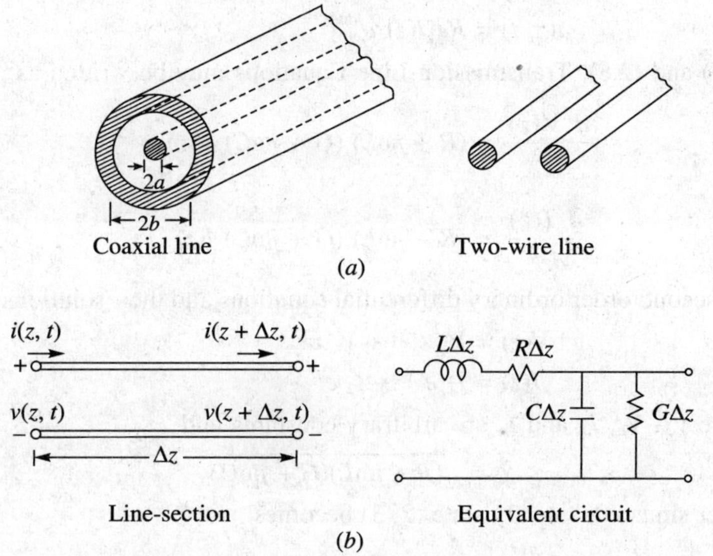

Fig. 2.1

The transmission line equivalent circuit shown in Fig. 2. 1(b) may be used to write relationships between currents and voltages at the two ends. Expanding $v(z + \Delta z, t)$ and $i(z + \Delta z, t)$ in Taylor series and ignoring the terms containing Δz^2 and higher powers, we have

$$v(z + \Delta z, t) = v(z, t) + \frac{\partial v(z, t)}{\partial z} \Delta z \qquad \text{...(2.2a)}$$

and
$$i(z + \Delta z, t) = i(z, t) + \frac{\partial i(z, i)}{\partial z} \Delta z \qquad \text{...(2.2b)}$$

Using Kirchhoff's laws and (2.2a) and (2.2b) we get

$$-\frac{\partial v(z, t)}{\partial z} = Ri(z, t) + L\frac{\partial i(z, t)}{\partial t} \qquad \text{...(2.3)}$$

$$-\frac{\partial i(z, t)}{\partial z} = Gv(z, t) + C\frac{\partial v(z, t)}{\partial t} \qquad \text{...(2.4)}$$

These two equations can be combined to yield two second order differential equations in v and i separately. These are

$$\frac{\partial^2 v}{\partial z^2} = RGv + (RC + LG)\frac{\partial v}{\partial t} + LC\frac{\partial^2 v}{\partial t^2} \qquad \text{...(2.5)}$$

and
$$\frac{\partial^2 i}{\partial z^2} = RGi + (RC + LG)\frac{\partial i}{\partial t} + LC\frac{\partial^2 i}{\partial t^2} \qquad \text{...(2.6)}$$

Equations (2.5) and (2.6) are known as Transmission Line Equations. For steady state sinusoidal excitation, voltage and current may be expressed as

$$v(z, t) = Re[V(z)\, e^{j\omega t}] \qquad ...(2.7)$$

$$i(z, t) = Re[I(z)\, e^{j\omega t}] \qquad ...(2.8)$$

Using (2.7) and (2.8), Transmission Line Equations may be written as

$$\frac{d^2V(z)}{dz^2} = (R + j\omega L)\,(G + j\omega C)\,V(z) \qquad ...(2.9)$$

$$\frac{d^2I(z)}{dz^2} = (R + j\omega L)\,(G + j\omega C)\,I(z) \qquad ...(2.10)$$

These are second order ordinary differential equations and their solutions are given by

$$V(z) = V_+\,e^{-\gamma z} + V_-\,e^{\gamma z} \qquad ...(2.11)$$

$$I(z) = I_+\,e^{-\gamma z} + I_-\,e^{\gamma z} \qquad ...(2.12)$$

where V_+, V_-, I_+ and I_- are arbitrary constants and

$$\gamma = \sqrt{(R + j\omega L)(G + j\omega C)} \qquad ...(2.13)$$

Under sinusoidal steady state (2. 3) becomes

$$-\frac{dV}{dz} = (R + j\omega L)\,I \qquad ...(2.14)$$

Substituting this in (2.11),

$$I = \sqrt{\frac{G + j\omega C}{R + j\omega L}}\,\{V_+e^{-\gamma z} - V_-e^{\gamma z}\} \qquad ...(2.15)$$

Defining $Z_0 = \sqrt{(R + j\omega L)/(G + j\omega C)} = (1/Y_0)$, equations 2.12 and 2.15 yield

$$I_+ = Y_0V_+ \text{ and } I_- = -Y_0V_- \qquad ...(2.16)$$

Z_0 is known as the *characteristic impedance* of the line and γ is called the *propagation constant.* Propagation constant is a complex quantity and may be written as

$$\gamma = \alpha + j\beta \qquad ...(2.17)$$

where α is known as the *attenuation constant* (expressed in nepers/unit length) and β the *phase constant* (expressed in radians/unit length). Z_0 and γ are the two parameters that characterize a uniform transmission line.

The two terms on right hand side of (2.11), namely $V_+\,e^{-\gamma z}$ and $V_-\,e^{\gamma z}$ represent two waves travelling in the positive z direction and the negative z direction, respectively. These waves are known as *incident wave* and *reflected wave,* respectively.

The constants V_+ and V_-, which are the amplitudes of the incident wave and the reflected wave, respectively, are evaluated using terminal conditions at the two ends of the line. Let us consider the terminal conditions shown in Fig. 2.2. Voltage and current along the line can be expressed either in terms V_s and I_s at $z = -s$ or in terms of voltage and current V_R and I_R at the other end ($z = 0$). Consider the second case. At $z = 0$, $V = V_R$ and $I = I_R$. Substituting these values in (2.11) and (2.15) we get

$$V_+ = (V_R + I_R Z_0)/2 \quad ...(2.18)$$

$$V_- = (V_R + I_R Z_0)/2 \quad ...(2.19)$$

Substituting for V_+ and V_- in (2.11) and (2.15), we get

$$V_s = V(-s) = V_R \cosh(\gamma s) + I_R Z_0 \sinh(\gamma s) \quad ...(2.20)$$

$$I_s = I(-s) = \frac{V_R}{Z_0} \sinh(\gamma s) + I_R \cosh(\gamma s) \quad ...(2.21)$$

These two equations express the voltage and the current at one end of the transmission line in terms of voltage and current at the other end. For network representation, *ABCD* parameters for a section of transmission line of length s may be written as

$$\begin{bmatrix} V_s \\ I_s \end{bmatrix} = \begin{bmatrix} A & B \\ C & D \end{bmatrix} \begin{bmatrix} V_R \\ I_R \end{bmatrix}$$

where

$$\begin{bmatrix} A & B \\ C & D \end{bmatrix} = \begin{bmatrix} \cosh(\gamma s), Z_0 \sinh(\gamma s) \\ Y_0 \sinh(\gamma s), \cosh(\gamma s) \end{bmatrix} \quad ...(2.22)$$

In the remaining part of this section on transmission lines, we shall consider lossless lines only. For lossless lines $R = G = 0$ (*see* Fig. 2.1) and γ may be written as $j\beta$. In most of the practical cases the line losses are small and this assumption is valid.

2.1.2 Reflection coefficient, standing waves and input impedance

Ratio of the amplitude of the reflected wave to that of the incident wave is called *reflection coefficient* ρ

$$\rho = \frac{V_-}{V_+} \quad ...(2.23)$$

For a lossless line $\alpha = 0$ and (2.11) and (2.15) may be written as

$$V(z) = V_+ e^{-j\beta z} (1 + \rho e^{2j\beta z}) \quad ...(2.24)$$

$$I(z) = Y_0 V_+ e^{-j\beta z} (1 - \rho e^{2j\beta z}) \quad ...(2.25)$$

Since z can be measured from any point, let us assume z to be zero at the load end (Fig. 2.2).

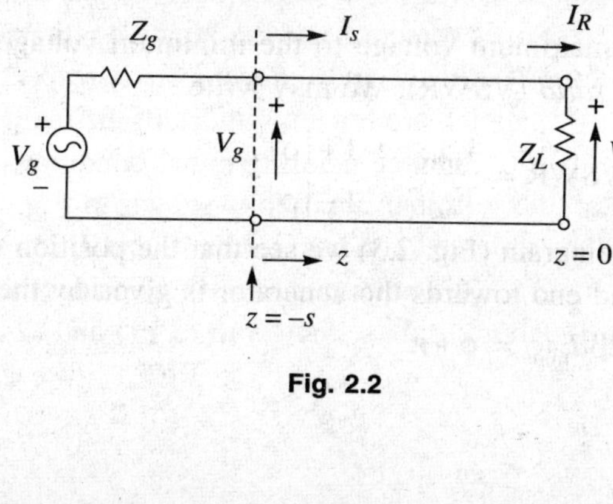

Fig. 2.2

At $z = 0$ (load end) we have $V_R = I_R Z_L$ and therefore (2.24) and (2.25) may be combined to yield

$$Z_L = Z_o \frac{1+\rho}{1-\rho} \quad \text{or} \quad \rho = \frac{Z_L - Z_0}{Z_L + Z_0} \qquad ...(2.26)$$

If we define a *generalized reflection coefficient* $\rho(z)$ as the ratio of reflected and incident components inclusive of propagation phases, that is,

$$\rho(z) = V_-(z)/V_+(z) = \rho e^{2J\beta z} \qquad ...(2.27)$$

equations 2.24 and 2.25 can be written as

$$V(z) = V_+ e^{-J\beta z}[1+\rho(z)] \qquad ...(2.24a)$$

$$I(z) = Y_0 V_+ e^{-J\beta z}[1-\rho(z)] \qquad ...(2.25a)$$

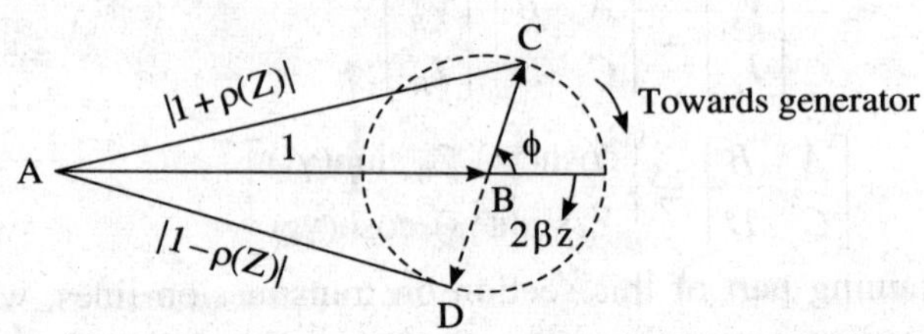

Fig. 2.3

Vector $V(z)/V_+$ may be represented as shown in Fig. 2.3. As one moves towards the generator, $\rho(z)$ rotates in clockwise direction. The minimum value of the magnitude of $V(z)/V_+$ is $1 - |\rho|$ and the maximum value is $1 + |\rho|$. Similarly, the variation of $I(z)/(Y_o V_+)$ may also be represented by a rotating vector. It can be easily seen that the position of the voltage maximum is the position of the current minimum and vice versa. Normalized voltage and current plotted along the line length are shown in Fig. 2.4(a).

For some simple cases standing wave patterns for voltages and currents are shown in Figures 2.4 (b), 2.4 (c) and 2.4 (d). When $Z_L = 0$, $\rho = -1$ and standing wave pattern is as shown in Fig. 2.4(b). For $Z_L = \infty$ ($\rho = 1$), Fig. 2.4(c) is applicable. When $Z_L = Z_0$, $\rho = 1$, there are no standing waves and the resulting pattern shown in Fig. 2.4 (d) is a straight line representing a uniform distribution of the voltage and the current along the line.

The ratio of the maximum voltage to the minimum voltage is defined as the *voltage standing wave ratio* (VSWR). We may write

$$\text{VSWR} = \frac{V_{max}}{V_{min}} = \frac{1+|\rho|}{1-|\rho|} \qquad ...(2.28)$$

From the vector diagram (Fig. 2.3) we see that the position of the first voltage minimum from the load end towards the generator is given by the relation

$$2\beta d_{min} = \phi + \pi \qquad ...(2.29)$$

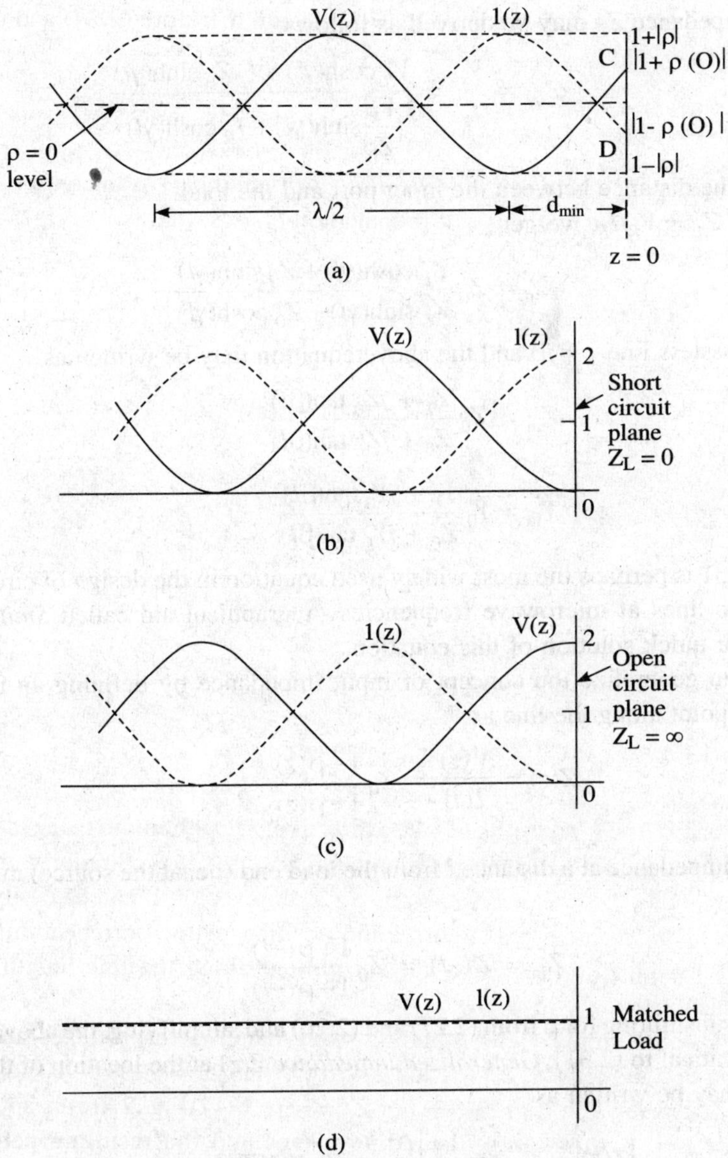

Fig. 2.4

where ϕ is the phase angle of ρ at the load end. Thus, by knowing the position of first voltage minimum, one can find the phase angle of p. Magnitude of p may be found by measuring VSWR and using (2.28).

Input impedance: Impedance looking into the two terminals at one end of the transmission line (when the other end is terminated in a known load impedance Z_L) is known as the input impedance Z_{in}. An expression for the input impedance in terms of the load impedance Z_L may be derived as follows:

$$Z_{in} = \frac{V_s}{I_s} = \frac{V_R \cosh(\gamma l) + I_R Z_0 \sinh(\gamma l)}{\frac{V_R}{Z_0}\sinh(\gamma l) + I_R \cosh(\gamma l)} \quad \text{...(2.30)}$$

where l is the distance between the input port and the load.

Since $Z_L = V_R/I_R$, we get

$$Z_{in} = Z_0 \frac{Z_L \cosh(\gamma l) + Z_0 \sinh(\gamma l)}{Z_L \sinh(\gamma l) + Z_0 \cosh(\gamma l)}$$

For lossless line $\gamma = j\beta$ and the above equation may be written as

$$Z_{in} = Z_0 \frac{Z_L + jZ_0 \tan(\beta l)}{Z_0 + jZ_L \tan(\beta l)} \quad \text{...(2.31)}$$

or

$$Y_{in} = Y_0 \frac{Y_L + jY_0 \tan(\beta l)}{Z_0 + jY_L \tan(\beta l)}$$

Equation 2.31 is perhaps the most widely used equation in the design of circuits using transmission lines at microwave frequencies. A graphical aid called *Smith chart* is available for quick solution of this equation.

We can generalize the concept of input impedance by defining an impedance $Z(z)$ at any point along the line as

$$Z(z) = \frac{V(z)}{I(z)} = Z_0 \frac{1+\rho(z)}{1-\rho(z)} \quad \text{...(2.32)}$$

Input impedance at a distance l from the load end (i.e. at the source) may now be written as

$$Z_{in} = Z(-l) = Z_0 \frac{1+\rho(-l)}{1-\rho(-l)} \quad \text{...(2.33)}$$

After substituting for ρ from (2.27) and (2.26) and simplifying, the above equation becomes identical to (2.31). *Generalized impedance $Z(z)$* at the location of the voltage maximum may be written as

$$|Z(z)|_{max} = Z_0 \frac{1+|\rho|}{1-|\rho|} = Z_0\, VSWR \quad \text{...(2.34)}$$

which is real and maximum. On the other hand the impedance at the location of voltage minimum is given by $Z_0/VSWR$ which is also real but has the minimum value.

2.1.3 Smith chart

As pointed out above, Smith chart is a graphical aid used for solving transmission line problems involving the use of equation (2.31) or (2.32) for Z_{in}. Writing $Z(z)/Z_0 = R + jX$ and $\rho(z) = u + jv$ we have

$$R + jX = \frac{1+u+jv}{1-u-jv} \quad \text{...(2.35)}$$

Equating real and imaginary parts on two sides of (2.35) we may write

$$\left(u - \frac{R}{1+R}\right)^2 + v^2 = \frac{1}{(1+R)^2} \qquad ...(2.36)$$

$$(u-1)^2 + \left(v - \frac{1}{X}\right)^2 = \frac{1}{X^2} \qquad ...(2.37)$$

Equations 2.36 and 2.37 represent two families of circles in complex $\rho (= u + jv)$ plane. The first set, given by (2.36), corresponds to circles for various values of parameter R. These circles have their centres at $(R/(1 + R), 0)$ and their radii are given by $1/(1 + R)$. These are called *constant resistance circles.* Equation 2.37, on the other hand, represents *constant reactance circles* with centres at $(1, 1/X)$ and radii given by $1/X$.

From 2.26 we note that for any positive Z_L, the maximum value of $|\rho|$ is unity. Thus, the area of interest in the complex (u, v) plane lies inside a circle with radius $|\rho|$ equal to unity. A graphical representation of this area with constant R and constant X circles as given by (2.36) and (2.37) shown therein, is known as *Smith chart.* The layout of the Smith chart is shown in Fig. 2.5. Commonly used charts have more constant R and constant X circles than shown here. Also, constant $|\rho|$ circles are

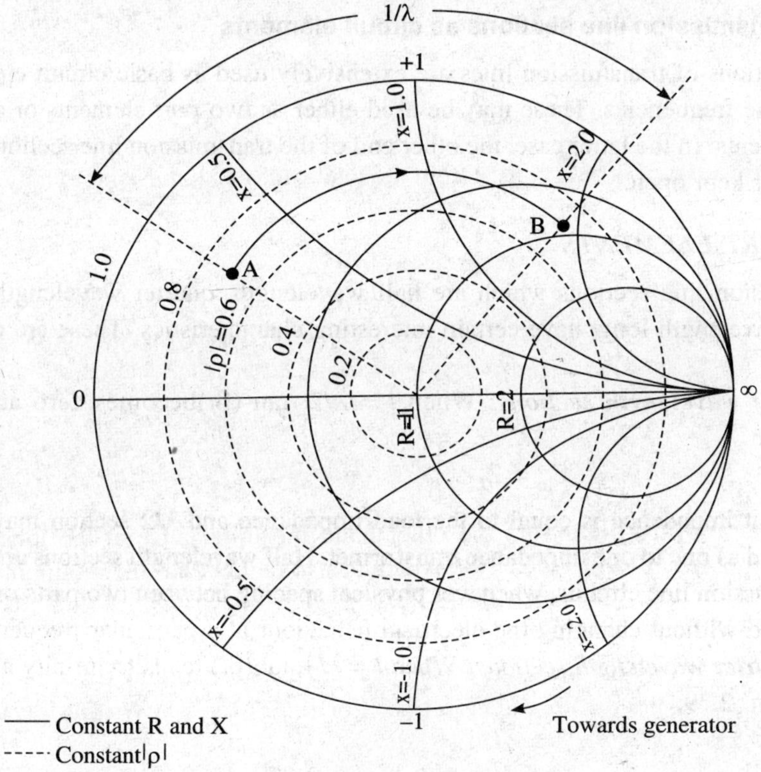

Fig. 2.5

normally not printed on the chart. Value of R and X as shown are normalized with respect to Z_0.

A few more observations about the Smith chart are in order. The point $R = 1$, $X = 0$ corresponds to $u = 0$, $v = 0$, the origin of the ρ plane. The distance from the origin to any other point gives magnitude of ρ and the angle between $X = 0$ line and the line connecting the point to the origin gives the phase angle of ρ. As already seen in the vector diagram (Fig. 2.3), when one moves along the transmission line from the load end towards the generator, the ρ vector moves clockwise. The information is noted on the Smith chart as 'towards generator'. Also, the angles $2\beta l (= 4\pi l/\lambda)$ are marked in terms of l/λ. The chart covers a range of l/λ equal to 0.5 which corresponds to $2\beta l = 2\pi$. For βl greater than π, $\rho(z)$ and $Z(z)$ repeat themselves.

In order to calculate Z_{in} for a given Z_L (i.e. to solve equation 2.31), one uses the fact that as we move along a uniform section of a transmission line, $|\rho|$ remains constant. Z_{in} is obtained as follows: (i) locate Z_L/Z_0 on the Smith chart, say point A in Fig. 2.5. (ii) draw a constant $|\rho|$ circle passing through A, (iii) move on this circle clockwise (towards generator) by the required distance l/λ from point A to point B and finally (iv) read R and X at point B to obtain Z_{in}/Z_0.

More familiarity with the applications of the Smith chart may be obtained by solving the problems listed at the end of this chapter.

2.1.4 Transmission line sections as circuit elements

Short sections of transmission lines are extensively used as basic circuit elements at microwave frequencies. These may be used either as two-port elements or as single-port elements. In the latter case, the other end of the transmission line section is either shorted or kept open.

TWO-PORT ELEMENTS

Transmission line sections which are half wavelength, quarter wavelength or one-eighth wavelength long, have certain interesting characteristics. These are discussed below.

Half wavelength sections: When $l = \lambda/2$, tan (βl)becomes zero and (2.31) yields

$$Z_{in} = Z_L \qquad ...(2.38)$$

Input impedance is equal to the load impedance and $\lambda/2$ section may thus be considered as one to one impedance transformer. Half wavelength sections are inserted in transmission line circuits, whenever physical spacing between two parts of a circuit is required without changing the electrical behaviour at a particular frequency.

Quarter wavelength sections: When $l = \lambda/4$, tan(βl) tends to infinity and (2.31) yields eqn. 2.39:

$$Z_{in} = Z_0^2 / Z_L \qquad ...(2.39)$$

This relation shows that if Z_L is resistive Z_{in} is also resistive but has a different value. Thus, a quarter wave section of a suitable impedance Z_0 may be used to transform a resistive impedance Z_L to another impedance value Z_{in}. When $Z_L = 0$, Z_{in} tends to infinity. This property can be used to provide high impedance stub supports for RF structures, as well as RF choke arrangements for DC bias in microwave circuits. Quarterwave sections (λ/4 long) are widely used for impedance transformation. As discussed later in Chapter 7, more than one λ/4 sections with different Z_0 may be cascaded together to increase the bandwidth of the impedance transformer.

One-eighth wavelength sections: When $L = l/8$, tan (βl)equals unity and (2.31) reduces to

$$Z_{in} = Z_0 \frac{Z_L + jZ_0}{Z_0 + jZ_L} \qquad ...(2.40)$$

That is, $|Z_{in}|$ equals Z_0 if Z_L is real. When Z_L is complex and $Z_0 = |Z_L|$, Z_{in} becomes real and is given by

$$Z_{in} = |Z_L| \sqrt{\left(\frac{R^2}{X} + X + |Z_L|\right) / \left(\frac{R^2}{X} + X - |Z_L|\right)} \qquad ...(2.41)$$

where R and X are resistive and reactive parts of Z_L. Thus, we note that a λ/8 section can be used to transform a complex impedance Z_L into a real impedance Z_{in} when $Z_0 = |Z_L|$.

SINGLE PORT ELEMENTS

Short circuited transmission line sections : When a section of transmission line is short circuited at the load end, $Z_L = 0$ and Z_{in} may be written as

$$Z_{in} = jZ_0 \tan(\beta l) \qquad ...(2.42)$$

which shows that Z_{in} is purely reactive. It is inductive for $\beta l < \pi/2$ (i.e. $l < \lambda/4$) and capacitive for $\pi/2 < \beta l < \pi$. (i.e. $\lambda/4 < l < \lambda/2$). Thus, a short circuited section of a transmission line may be used as an inductive reactance or as a capacitive reactance by suitable choice of its length.

Open circuited transmission line sections: When a section of a transmission line is open circuited at the load end, Z_L tends to infinity and Z_{in} may be written as

$$Z_{in} = -jZ_0 \cot(\beta l) \qquad ...(2.43)$$

Again Z_{in} is pure reactance. However, now it is capacitive for $\beta l < \pi/2$ and inductive for $\pi/2 < \beta l < \pi$. Thus, an open circuited section may also be used as an inductive or capacitive reactance. Since it is often more convenient to have ideal short circuit termination rather than an ideal open circuit termination (particularly in the case of coaxial lines), short circuited sections of transmission lines find more frequent use.

2.2 WAVEGUIDES

Waveguides provide an alternative to transmission lines for use at microwave frequencies. In fact, waveguides are superior to conventional transmission lines in terms of attenuation per unit length experienced by the wave propagating through them. In this section we will review two most commonly used forms of waveguides. These are rectangular and circular cylindrical waveguides shown in Fig. 2.6. Waveguide analysis is based on the solution of wave equation for electromagnetic fields.

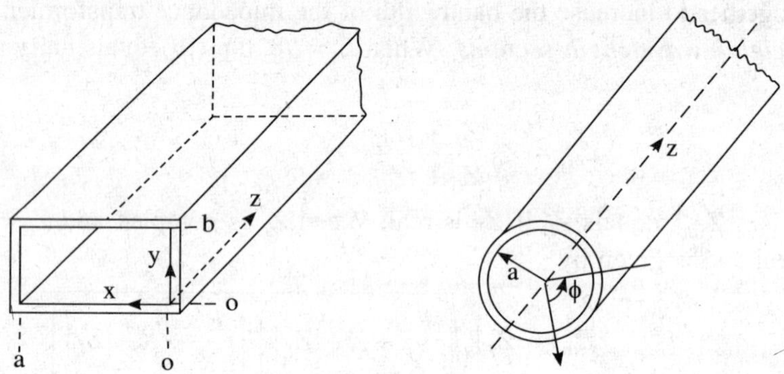

Fig. 2.6

2.2.1 Wave equation

Waveguides are analysed in terms of electric and magnetic fields inside the guide. Basic laws that govern the behaviour of these fields are Maxwell's equations which may be written as

$$\nabla \times \tilde{\mathbf{H}} = \partial \tilde{\mathbf{D}} / \partial t + \tilde{\mathbf{J}} \quad \text{...(2.44)}$$

$$\nabla \times \tilde{\mathbf{E}} = -\partial \tilde{\mathbf{B}} / \partial t \quad \text{...(2.45)}$$

$$\nabla \cdot \tilde{\mathbf{D}} = \rho \quad \text{...(2.46)}$$

$$\nabla \cdot \tilde{\mathbf{B}} = 0 \quad \text{...(2.47)}$$

Vectors $\tilde{\mathbf{E}}$ and $\tilde{\mathbf{H}}$ denote electric and magnetic fields, and $\tilde{\mathbf{D}}(=\varepsilon\tilde{\mathbf{E}})$ and $\tilde{\mathbf{B}}(=\mu\tilde{\mathbf{H}})$ are electric and magnetic flux densities, respectively.* Constants ε and μ represent the permittivity and the permeability, respectively, of the medium and **J** and ρ are electric current density and electric charge density, respectively. For perfect dielectric media containing no charges and no conduction currents, one can substitute $\rho = 0$ and $\tilde{\mathbf{J}} = 0$. Taking curl of (2.45) and substituting (2.44), we have for $\tilde{\mathbf{J}} = 0$

$$\nabla \times \nabla \times \tilde{\mathbf{E}} = -\mu\varepsilon \partial^2 \mathbf{E} / \partial t^2 \quad \text{...(2.48)}$$

Using vector identity $\nabla \times \nabla \times \mathbf{A} = (\nabla\nabla \cdot \mathbf{A} - \nabla^2 \mathbf{A})$

write $\nabla\nabla \cdot \tilde{\mathbf{E}} - \nabla^2 \tilde{\mathbf{E}} = -\mu\varepsilon \partial^2 \mathbf{E} / \partial t^2$ ...(2.49)

---

* Superscript ~ indicates that all these vectors are arbitrary functions of time.

Since ρ = 0, we have $\nabla \cdot \tilde{\mathbf{D}}$ and $\nabla \cdot \tilde{\mathbf{E}}$ equal to zero and (2.49) becomes

$$\nabla^2 \tilde{\mathbf{E}} = \mu\varepsilon \partial^2 \tilde{\mathbf{E}}/\partial t^2 \qquad ...(2.50)$$

which is the wave equation for the electric field vector $\tilde{\mathbf{E}}$.
We can derive a similar equation for the magnetic field vector $\tilde{\mathbf{H}}$.

$$\nabla^2 \tilde{\mathbf{H}} = \mu\varepsilon \partial^2 \tilde{\mathbf{H}}/\partial t^2 \qquad ...(2.51)$$

The time variation of $\tilde{\mathbf{E}}$ is expressed by $e^{j\omega t}$, i.e.

$$\tilde{\mathbf{E}}(r,t) = Re\{\mathbf{E}(r)e^{j\omega t}\} \qquad ...(2.52)$$

where ω is angular, frequency in radians. $\mathbf{E}(r)$ is a complex vector and is not a function of time. Other variables also follow a similar variation. Equation 2.50 can be written as

$$\nabla^2 \mathbf{E} = -\omega^2 \mu\varepsilon \mathbf{E} \qquad ...(2.53)$$

The solution of the wave equation 2.53, subject to the appropriate boundary conditions at the waveguide walls, leads to the transmission characteristics of the waveguides. The relevant boundary conditions on the conducting walls are: the tangential component of **E** field should be zero and the normal component of **H** field be zero. Solution of (2.53) for two types of waveguides, rectangular and circular, are discussed separately.

2.2.2 Rectangular waveguides

Consider the rectangular waveguide geometry shown in Fig. 2.6. Assuming that the variation in z-direction map be expressed as $e^{-\gamma z}$ where $\gamma = \alpha + j\beta$, Maxwell's equations 2.44 and 2.45 become **J** = ρ = 0 within the guide).

$$\left.\begin{aligned} \frac{\partial H_z}{\partial y} + \gamma H_y &= j\omega\varepsilon\, E_x \\ \frac{\partial H_z}{\partial x} + \gamma H_x &= -j\omega\varepsilon\, E_y \\ \frac{\partial H_y}{\partial x} - \frac{\partial H_x}{\partial y} &= j\omega\varepsilon\, E_x \end{aligned}\right\} \qquad ...(2.54)$$

and

$$\left.\begin{aligned} \frac{\partial E_z}{\partial y} + \gamma E_y &= -j\omega\mu\, H_x \\ \frac{\partial E_z}{\partial x} + \gamma E_x &= j\omega\mu\, H_y \\ \frac{\partial E_y}{\partial x} - \frac{\partial E_x}{\partial y} &= -j\omega\mu\, H_z \end{aligned}\right\} \qquad ...(2.55)$$

The wave equations for E_z and H_z which have been derived earlier, may be written as

$$\frac{\partial^2 E_z}{\partial x^2}+\frac{\partial^2 E_2}{\partial y^2}+\gamma^2 E_z = -\omega^2\mu\varepsilon E_z \qquad ...(2.56)$$

$$\frac{\partial^2 H_z}{\partial x^2}+\frac{\partial^2 H_2}{\partial y^2}+\gamma^2 H_z = -\omega^2\mu\varepsilon H_z \qquad ...(2.57)$$

Equations 2.54 and 2.55 can be obtained to yield

$$\left.\begin{aligned} H_x &= -\frac{\gamma}{h^2}\frac{\partial H_z}{\partial x}+j\frac{\omega\varepsilon}{h^2}\frac{\partial E_z}{\partial y} \\ H_y &= -\frac{\gamma}{h^2}\frac{\partial H_z}{\partial y}-j\frac{\omega\varepsilon}{h^2}\frac{\partial E_z}{\partial x} \end{aligned}\right\} \qquad ...(2.58)$$

$$\left.\begin{aligned} E_x &= -\frac{\gamma}{h^2}\frac{\partial E_z}{\partial x}-j\frac{\omega\mu}{h^2}\frac{\partial H_z}{\partial y} \qquad ...(2.59) \\ E_y &= -\frac{\gamma}{h^2}\frac{\partial E_z}{\partial y}+j\frac{\omega\mu}{h^2}\frac{\partial H_z}{\partial x} \qquad ...(2.59) \end{aligned}\right\}$$

where $h_2 = \gamma^2+\omega^2\mu\varepsilon$ ...(2.60)

Equations 2.58 and 2.59 show that all other field components can be expressed in terms of E_z and H_z. Thus if E_z and H_z are both zero, all the fields within the guide will vanish. Since either an E_z or an H_z component must exist, it is convenient to divide the possible field configurations within the guide into two sets; ***transverse magnetic*** (TM) ***waves* for which $H_z = 0$, and *transverse electric*** (TE) ***waves* for which** $E_z = 0$. In **general, the waveguide field may be a combination** of these **types of waves.**

TM waves : For TM waves H_z is zero and we should solve (2.56) for E_z subject to the following boundary conditions

$$E_z = 0 \text{ at } x = 0,\ x = a,\ y = 0 \text{ and } y = b \qquad ...(2.61)$$

The **partial differential equation** (2.56) can be solved by usual technique of **assuming** a **product solution** of the **type** $E_z = XY\,e^{-\gamma z}$ and obtaining two ordinary **differential** equations in x and y **separately.** We get

$$E_z = (C_1 \cos Bx + C_2 \sin Bx)(C_3 \cos Ay + C_4 \sin Ay)\,e^{-\gamma z} \quad (2.62)$$

where $B^2 = h^2 - A^2$. The **constants** C_1, C_2, C_3, C_4, A and B are to be selected to satisfy the boundary conditions (2.61). We obtain

$$\begin{aligned} C_1 &= 0 \\ C_3 &= 0 \\ A &= n\pi/b \\ B &= m\pi/a \end{aligned} \qquad ...(2.63)$$

Using (2.63), the resultant expression for E_z may be written as

$$E_z = C\sin\left(\frac{m\pi}{a}x\right)\sin\left(\frac{n\pi}{b}y\right)e^{-\gamma z} \qquad ...(2.64)$$

where $C = C_2C_4$. Other field components may be evaluated using (2.58) and (2.59).

Since $A^2 + B^2 = h^2$, (2.60) may be written as

$$\gamma = \sqrt{A^2 + B^2 - \omega^2\mu\varepsilon}$$

$$= \sqrt{\left(\frac{m\pi}{a}\right)^2 + \left(\frac{n\pi}{b}\right)^2 - \omega^2\mu\varepsilon} \qquad ...(2.65)$$

Equation 2.65 determines the propagation constant for TM waves in a rectangular waveguide. At low frequencies, $\omega^2\mu\varepsilon$ is small and γ is real. Since $\gamma = \alpha + j\beta$ where α is the attenuation constant and β the phase constant, γ being real implies that $\beta = 0$, i.e. there is no transmission of waves at these frequencies. When $\omega^2\mu\varepsilon$ is greater than $[(m\pi/a)^2 + (n\pi/b)^2]$, γ will be imaginary and may be written as $j\beta$. In this frequency range the attenuation constant is zero, if the conductor and the dielectric losses in the guide are neglected.

The frequency below which the wave propagation does not take place is called *cut-off frequency* and is given by

$$\omega_c = \frac{1}{\sqrt{\mu\varepsilon}}\sqrt{\left(\frac{m\pi}{a}\right)^2 + \left(\frac{n\pi}{b}\right)^2} \qquad ...(2.66)$$

Above the cut-off frequency, phase constant β can be written as

$$\beta = \sqrt{\omega^2\mu\varepsilon - \omega_c^2\mu\varepsilon} \qquad ...(2.67)$$

and the phase velocity of wave propagation as

$$v = \frac{\omega}{\beta} = \frac{1}{\sqrt{\mu\varepsilon}\,\sqrt{1-\omega_c^2/\omega^2}} = \frac{c}{\sqrt{1-\omega_c^2/\omega^2}} \qquad ...(2.68)$$

where c (equal to $1/\sqrt{\mu\varepsilon}$ is the velocity of waves in free space.

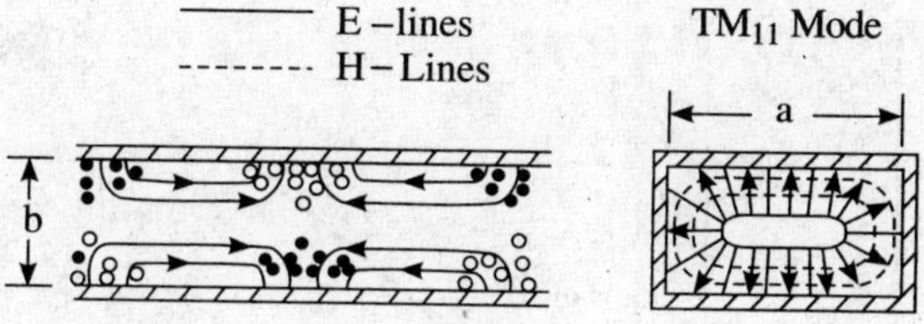

Fig. 2.7

The above result indicates that the velocity of propagation of the wave in the guide is greater than the phase velocity in free space. As the frequency is increased above the cut-off value ω_c, the phase velocity decreases from an infinitely large value and approaches the velocity in free space. Wavelength within the guide is called guide-wavelength λ_g and is given by

$$\lambda_g = \frac{2\pi}{\sqrt{(\omega^2\mu\varepsilon - \omega_c^2\mu\varepsilon)}} \qquad ...(2.69)$$

Since m and n can have any integral value, multiple solutions for phase constant β are possible. These are called *modes of propagation.* These transverse magnetic modes are designated as TM_{mn} modes. As can be seen from (2.64), E_z becomes zero when either m or n is zero. So the cut-off frequency given by (2.66) would be minimum when $m = n = 1$. This mode is called TM_{11} mode. Distribution of E and H fields for this mode is shown in Fig. 2.7.

TE waves: Characteristics of transverse electric (TE) waves can be derived in a manner similar to that used for TM waves. Now $E_z = 0$ and the solution for H_z is found to be of the same general form as (2.62). For evaluation of the constants, H_z is differentiated with respect to x and y to obtain E_x and E_y and the boundary conditions are then applied to E_x and E_y. H_z is thus found to be

$$H_z = C \cos (Bx) \cos (Ay)\, e^{-\gamma z} \qquad ...(2.70)$$

Expressions for γ, ω_c v and λ_g are identical to those for TM waves given in equations 2.65 to 2.69.

For TE_{mn} waves it is possible to make either m or n (but not both)

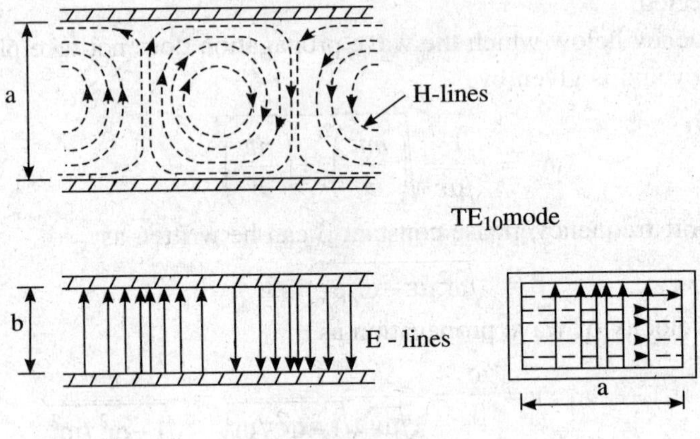

Fig. 2.8

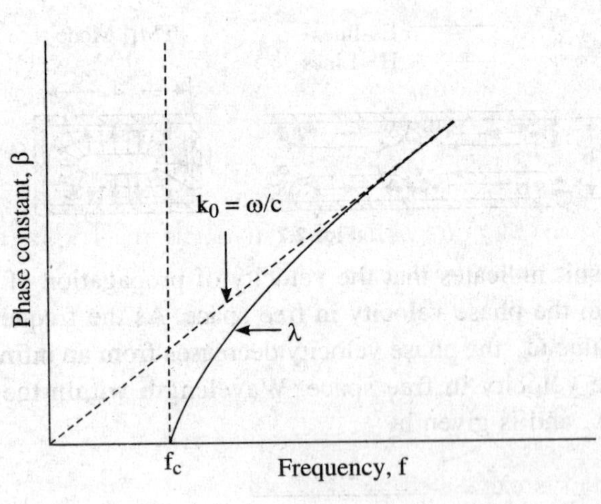

Fig. 2.9

equal to zero without causing all fields to vanish. It is seen from (2.66) that the lowest order mode (TE or TM) in a rectangular waveguide is the TE_{10} mode. This mode has the lowest cut-off frequency and is called the *dominant mode.*

TE_{10} mode is the most commonly used mode in rectangular waveguides. Field components for this mode ($m = 1$ and $n = 0$) may be written as

$$H_z = C\cos\left(\frac{\pi x}{a}\right)$$

$$H_x = \frac{j\beta aC}{\pi}\sin\left(\frac{\pi x}{a}\right)$$

$$E_y = \frac{-j\omega\mu aC}{\pi}\sin\left(\frac{\pi x}{a}\right) \qquad \text{...(2.71)}$$

$$E_x = H_y = 0$$

Variations with z and t are not shown in (2.71). Also for TE_{10} mode we have

$$\omega_c = \frac{1}{\sqrt{\mu\varepsilon}}\frac{\pi}{a} = \frac{c\pi}{a}$$

$$\beta = \sqrt{\omega^2\mu\varepsilon - (\pi/a)^2} \qquad \text{...(2.72)}$$

Cut-off wavelength, i.e. the free space wavelength corresponding to the cut-off frequency is given by $\lambda_c = 2a$. E and H field distributions for TE_{10} mode are shown in Fig. 2.8. Figure 2.9 shows the variation of β for this mode.

ATTENUATION IN WAVEGUIDES

In the waveguide analysis discussed above, it has been assumed that the walls are perfect conductors and the dielectric medium is lossless. Thus for frequencies greater than the cut-off frequency ω_c, propagation constant γ is purely imaginary (as given by eqn. 2.65) and the wave propagates unattenuated. However, in practical waveguides, losses are present mainly because of the finite conductivity of conducting walls.

As long as the losses are small, the following procedure may be used for calculating the attenuation constant. In this approach, it is assumed that the finite conductivity walls has only a small effect on the field configuration within the guide. In particular, the magnetic field tangential to the wall is expected to depend only slightly on the conductivity of walls. Thus the tangential magnetic field strength computed for perfectly conducting walls is used to determine the linear current density in the walls. This linear current density, squared and multiplied by the actual surface resistance of the walls, gives the actual power loss per unit area in the walls. The attenuation factor is thus given by

$$\alpha = \frac{\text{Power lost per unit length}}{2(\text{Power transmitted})} \qquad \text{...(2.73a)}$$

The power transmitted is calculated by integrating the z-component of the Poynting vector over the cross section of the guide, i.e. for the dominant mode

$$P_z = -\frac{1}{2}\int (E_y H_x^*) dx\, dy \qquad ...(2.73b)$$

$$= \frac{1}{2}\frac{\omega\mu\beta a^2 C^2}{\pi^2}(ab/2)$$

where C is a constant used in eqn. (2.71).

The power lost per unit length of the guide is

$$W_{lost} = \frac{1}{2} R_s \int_s |J|^2\, da \qquad ...(2.73c)$$

$$= \frac{1}{2} R_s \int_s |H_{tan}|^2\, da$$

where s denotes the surface area of guide walls and R_s is the surface resistivity, i.e. the resistance of the plane conductor for a unit length and unit width. Therefore, from (2.73a), (2.73b) and (2.73c),

$$\alpha = \frac{R_s}{b\eta_0}\frac{k_0}{\beta}\left[1 + 2\frac{b}{a}(f_c/f)^2\right] \qquad ...(2.73d)$$

where $\eta_0 = \sqrt{\mu_0/\varepsilon_0}$ is the intrinsic impedance of the free space.

A typical variation of α with frequency is shown in Fig. 2.10. It may be noted that the attenuation becomes infinite at the cut-off frequency, decreases to a minimum when frequency is increased and then increases gradually and monotonically as the frequency is increased further.

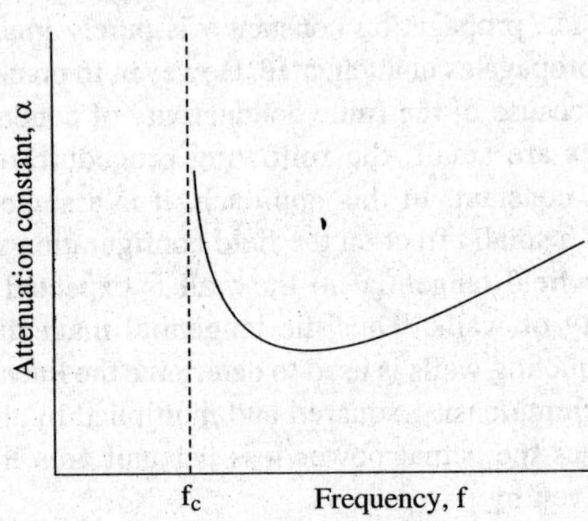

Fig. 2.10

Waveguides are used in the frequency range where attenuation is minimum. An idea of the typical numerical values can be obtained by considering the example given in Problems 2.10 and 2.11 at the end of this chapter.

EXCITATION OF WAVEGUIDES

In any practical set-up it often becomes necessary to feed a waveguide from a coaxial line or another waveguide. The method of excitation is selected such that the desired mode is excited and maximum power transfer takes place into the waveguide. A probe (quite often the central conductor of the transmission line) extending into the waveguide is the most common method of excitation. The location of the probe is chosen to produce lines of E and H fields that are roughly parallel to lines of E and H fields for the mode to be excited. For TE_{10} (i.e. the dominant) mode the probe location is shown in Fig. 2.11 (a). Since the E field component for this mode is along y and is maximum at $x = a/2$, the probe is located in the middle of the broad wall. On the other hand, if it were desired to excite TM_{11} mode, the appropriate orientation of the probe will be in z-direction as shown in Fig. 2.11(b). Similarity between the magnetic field produced by the probe to the H field distributions of respective modes can be observed by comparing Fig. 2.8 with Fig. 2.11 (a), and Fig. 2.7 with Fig. 2.11 (b).

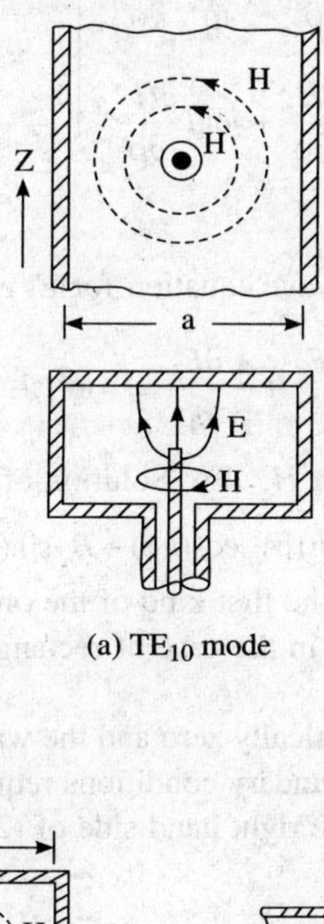

(a) TE_{10} mode

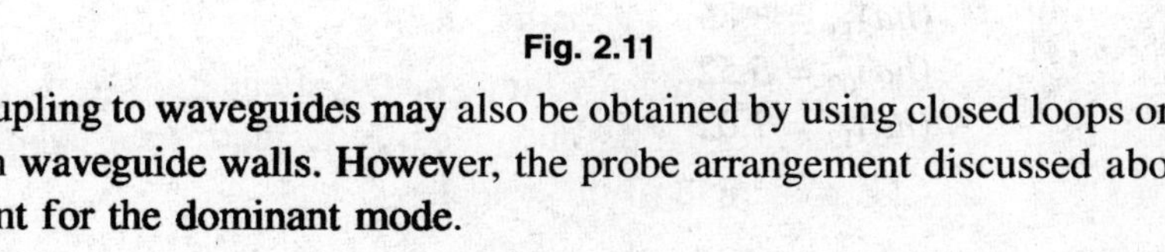

(b) TM_{11} mode

Fig. 2.11

Coupling to waveguides may also be obtained by using closed loops or apertures (holes) in waveguide walls. However, the probe arrangement discussed above is very convenient for the dominant mode.

2.2.3 Circular waveguides

Analysis of circular waveguides (shown in Fig. 2.6) requires solution of the wave equation in cylindrical coordinates. This simplifies application of boundary conditions for tangential components of the electric field on the cylindrical wall. Procedure is similar to that followed for rectangular waveguides. We express Maxwell's equations in cylindrical coordinates and then rearrange them to yield H_ρ, H_ϕ and E_ρ, E_ϕ in terms of two longitudinal components E_z and H_z. These relations are as follows:

$$\left.\begin{aligned} h^2 H_\rho &= j\frac{\omega\varepsilon}{\rho}\frac{\partial E_z}{\partial\phi} - \gamma\frac{\partial H_z}{\partial\rho} \\ h^2 H_\phi &= -j\omega\epsilon\frac{\partial E_z}{\partial\rho} - \frac{\gamma}{\rho}\frac{\partial H_z}{\partial\phi} \\ h^2 E_\rho &= -\gamma\frac{\partial E_z}{\partial\rho} - j\frac{\omega\mu}{\rho}\frac{\partial H_z}{\partial\phi} \\ h^2 E_\phi &= -\frac{\gamma}{\rho}\frac{\partial E_z}{\partial\phi} + j\omega\mu\frac{\partial H_z}{\partial\rho} \end{aligned}\right\} \quad \text{...(2.74)}$$

where $h^2 = \gamma^2 + \omega^2\mu\varepsilon$.

In cylindrical coordinates the wave equation for E_z is given by

$$\frac{\partial^2 E_z}{\partial\rho^2} + \frac{1}{\rho^2}\frac{\partial^2 E_z}{\partial\phi^2} + \frac{\partial^2 E_z}{\partial z^2} + \frac{1}{\rho}\frac{\partial E_z}{\partial\rho} = -\omega^2\mu\epsilon\, E_z \quad \text{...(2.75)}$$

A similar equation can be derived for H_z also. Solution of (2.75) may be written as

$$E_z = J_n(\rho h)\{A_n \cos(n\phi) + B_n \sin(n\phi)\}e^{-yz} \quad \text{...(2.76)}$$

where $J_n(\rho h)$ is Bessel's function of the first kind of the order n. Solution for H_z will also have exactly the same form. As in the case of rectangular waveguides, TM and TE modes are discussed separately.

TM waves **: For TM waves H_z is** identically zero and the wave equation for E_z is used **for carrying out the analysis. The** boundary conditions require that E_z must vanish at the surface of the guide. Equating the right hand side of (2.76) to zero for $\rho = a$, we have

$$j_n\,(ha) = 0 \quad \text{...(2.77)}$$

This shows that infinite number of modes corresponding to the solutions of (2.77) are possible. The first few modes are obtained when

$$\begin{aligned} (ha)_{01} &= 2.405 \\ (ha)_{11} &= 3.85 \\ (ha)_{02} &= 5.52 \\ (ha)_{12} &= 7.02 \end{aligned} \quad \text{...(2.78)}$$

The first subscript refers to the value of n and the second to the roots for a given n in increasing order of magnitude. TM modes corresponding to these solutions are referred to as TM_{01}, TM_{11}, TM_{02} and TM_{12}, respectively. For these modes we can write

$$\beta_{nm} = \sqrt{\omega^2\mu\varepsilon - h_{nm}^2} \qquad \text{...(2.79a)}$$

$$\omega_c = h_{nm} / \sqrt{\mu\varepsilon} \qquad \text{...(2.79b)}$$

$$v = \frac{\omega}{\beta} = \frac{\omega}{\sqrt{(\omega^2\mu\varepsilon - h_{nm}^2)}} \qquad \text{...(2.79c)}$$

where $h_{nm} = (ha)_{nm} / a$.

TE waves : For TE waves the longitudinal electric field component E_z is identically zero and H_z is given by an equation similar to (2.76). For applying boundary conditions in this case it is necessary to evaluate E_ϕ and put E_ϕ equal to zero at $\rho = a$. Thus E_ϕ is obtained as

$$E_\phi = \frac{j\omega\mu C_n}{h} J'n(h\rho)\cos(n\phi)e^{-\gamma z} \qquad \text{...(2.80)}$$

where the prime in $J'_n(h\rho)$ indicates the derivative with respect to the argument ($h\rho$). The boundary condition $E_\phi = 0$ at $\rho = a$ becomes

$$J'_n(ha) = 0 \qquad \text{...(2.81)}$$

The first few roots of this equation are written as

$$\begin{aligned} (ha)'_{11} &= 1.84 \\ (ha)'_{01} &= 3.83 \\ (ha)'_{12} &= 5.33 \\ (ha)'_{02} &= 7.02 \end{aligned} \qquad \text{...(2.82)}$$

The corresponding TE modes are referred to as TE_{11}, TE_{01}, and so on. Equations (2.79) hold good for these modes also; h_{nm} in this case being $(ha)'_{nm} / a$.

Comparing the magnitudes of roots given in (2.78) and (2.82) we find that for circular waveguides the mode having the lowest cut-off is TE_{11}, the next higher being TM_{01}. Field configurations for these two modes are shown in Fig. 2.12.

2.3 RESONATORS

2.3.1 Transmission line resonators

It was pointed out in Sec. 2.1.4 that short sections of transmission lines shorted at one end may be used to provide inductive and capacitive reactances by suitable selection of their lengths. Inductive and capacitive reactances thus obtained may be combined together to obtain a distributed circuit equivalent of a parallel resonant *L-C* circuit.

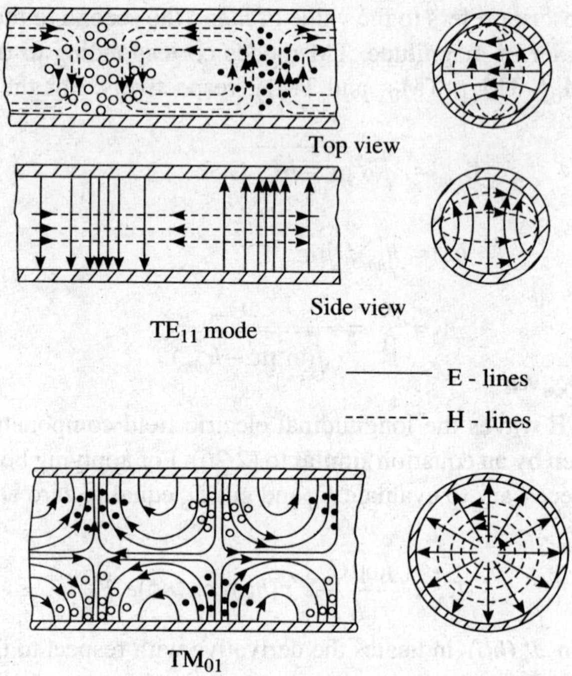

Fig. 2.12

Consider a shorted transmission line of length l_1 (Fig. 2.13). When l_1 is less than $\lambda/4$ the input susceptance is inductive and is given by

$$B_L = Y_0 \cot \beta l_1 \qquad ...(2.83)$$

For resonance, a capacitive susceptance B_C (= – B_L) is needed. This can be obtained by another shorted line length l_2 such that

$$B_C = -Y_0 \cot \beta l_1 = Y_0 \cot(\pi - \beta l_1) = Y_0 \cot \beta l_2 \qquad ...(2.84)$$

which yields

$$\beta l_2 = \pi - \beta l_1, \text{ or } \beta(l_1 + l_2) = \pi, \text{ or } (l_1 + l_2) = \lambda/2 \qquad ...(2.85)$$

This indicates that if the length of a transmission line is $\lambda/2$ and short circuits are placed at both the ends, the resulting structure will behave like a resonant circuit at the frequency where its length is $\lambda/2$. Similar to any *L-C* resonator, the transmission line resonator is capable of storing electromagnetic energy. In a lumped element *L-C* resonator the energy is stored in the magnetic field of the inductor and the electric field of the capacitor alternately. For a transmission line resonator, the electric and magnetic fields are distributed all over the resonator and the energy storage is not localized.

Resonators of the type discussed above could also be designed by using waveguide sections in place of the transmission lines. In order to visualize the reactances provided by waveguide sections, it is necessary to develop the concept of 'characteristic impedance' for waveguides. This is discussed in detail in Chapter 5. Presently, in this section, we will extend the technique of waveguide analysis for studying the waveguide resonators shown in Fig. 2.14.

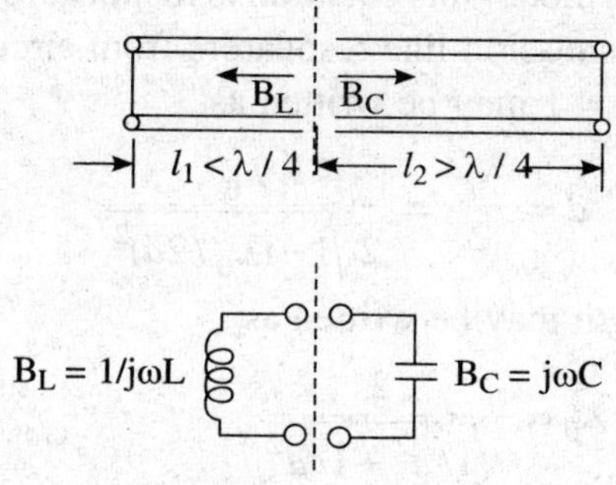

Fig. 2.13

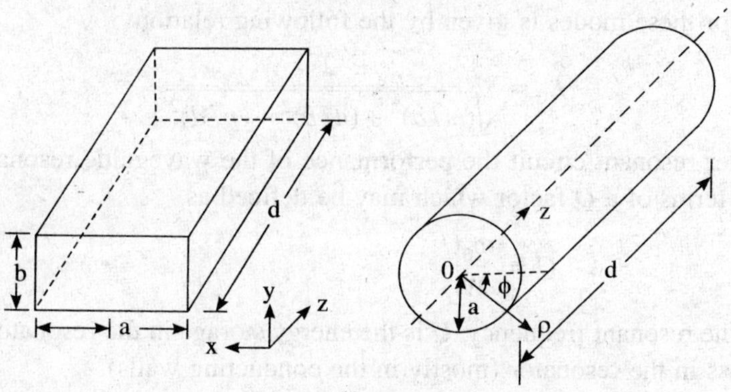

Fig.2.14

2.3.2 Rectangular waveguide resonator

We consider the dominant TE_{10} mode in the waveguide. The length of the waveguide is 'd' and two conducting walls are located at $z = 0$ and $z = d$. The field distribution for H_z now consists of two waves, one travelling in $+ z$ direction and the other travelling in $- z$ direction. H_z may therefore be expressed as

$$H_z = (Ce^{-j\beta z} + De^{j\beta z}) \cos \frac{\pi x}{a} \qquad ...(2.86)$$

Thus, expression for H_z must satisfy boundary conditions at $z = 0$ and $z = d$, namely

$$H_z = 0 \text{ at } z = 0 \text{ and } z = d \qquad ...(2.87)$$

Therefore

$$C = - D \text{ and } \beta d = \pi p \text{ (p = 1, 2, ...)} \qquad ...(2.88)$$

This shows that for electromagnetic fields to exist in the enclosed waveguide structure shown in Fig. 2.14, the dimension d should be chosen so that

$$d = p\frac{\lambda_g}{2} \quad \text{...(2.89)}$$

For $p = 1$ (lowest resonator mode) this equation is identical to the resonance condition (2.85) derived for the transmission line resonators from circuit considerations. Value of d given by (2.89) for $p = 1$ may be written as

$$d = \frac{\lambda_g}{2} = \frac{\lambda_0}{2\sqrt{1-(\lambda_0/2a)^2}} \quad \text{...(2.90)}$$

and the resonant wavelength may be written as

$$\lambda_0 = \frac{2}{\sqrt{1/a^2 + 1/d^2}} \quad \text{...(2.91)}$$

The dominant resonator mode considered above is known as TE_{101} mode.* Higher order modes are also possible and are designated as TE_{mnp} modes. The resonant wavelength for these modes is given by the following relation

$$\lambda_0 = \frac{2}{\sqrt{(m/a)^2 + (n/b)^2 + (p/d)^2}} \quad \text{...(2.92)}$$

Like any other resonant circuit the performance of the waveguide resonators is also expressed in terms of a Q factor which may be defined as

$$Q = \frac{\omega_0 U}{W_L} \quad \text{...(2.93)}$$

where ω_0 is the resonant frequency, U is the energy storage in the resonator and W_L is the power loss in the resonator (mostly in the conducting walls).

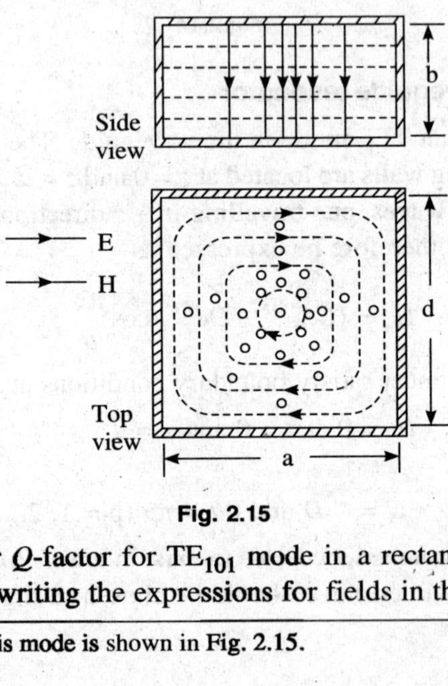

Fig. 2.15

An expression for Q-factor for TE_{101} mode in a rectangular resonator may be derived as follows. By writing the expressions for fields in the cavity we find that in

* Field configuration for this mode is shown in Fig. 2.15.

a rectangular cavity (as also in resonators of other shapes) E and H fields are out of phase by 90°. So, at a given instant when the electric field is maximum, the magnetic field is zero. The stored energy gets converted from the electric energy to the magnetic energy and vice versa. We may, therefore, calculate the energy storage at the instant when electric field is maximum and write

$$U = (U_E)_{max} = \frac{\varepsilon}{2}\int_0^d \int_0^b \int_0^a |E_y|^2 \, dxdydz \qquad ...(2.94)$$

where

$$E_y = E_0 \sin(\pi x / a)\sin(\pi z / d) \text{ and } E_0 = (2\omega\mu a C/\pi)$$

Carrying out the integration we get

$$U = \varepsilon abdE_0^2 / 8 \qquad ...(2.95)$$

To obtain an expression for power loss in the walls, we calculate the currents flowing in the walls from the values of the tangential components of the magnetic field at the walls. (These magnetic field components are calculated assuming the walls to be perfect conductors.) If the conducting walls have a surface resistivity R_s, the foregoing currents will produce a power loss given by

$$W_L = \frac{R_s}{2}\left\{2\int_0^b \int_0^a |H_x|^2_{z=0} dx\, dy + 2\int_0^d \int_0^b |H_z|^2_{x=0} dy\, dz \right.$$

$$\left. +2\int_0^d \int_0^a \left[|H_x|^2 + |H_z|^2\right]\Big|_{y=0} dx\, dz \right\} \qquad ...(2.96)$$

In this expression, the first term accounts for the losses in the front and the back walls, the second term for the losses in the left and the right sides and the third term for those in the top and the bottom surfaces. Substituting the expressions for H_x and H_z and putting $\beta = \pi/d$ we get the total power loss as

$$W_L = \frac{R_s\lambda_0^2}{8\eta^2} E_0^2 \left\{\frac{ab}{d^2} + \frac{bd}{a^2} + \frac{1}{2}\left(\frac{a}{d} + \frac{d}{a}\right)\right\} \qquad ...(2.97)$$

where $\eta = \sqrt{\mu/\varepsilon}$.

From (2.95) and (2.97), Q of the TE_{101} mode rectangular cavity may be written as

$$Q = \frac{\omega_0\sqrt{\mu\varepsilon}\eta}{R_s\lambda_0^2}\left\{\frac{2ba^3d^3}{2b(a^3 + d^3) + ad(a^2 + d^2)}\right\} \qquad ...(2.98a)$$

Substituting for the resonant wavelength λ_0 from (2.91), we get

$$Q = \frac{\pi\eta}{4R_s}\left\{\frac{2b(a^2 + d^2)^{3/2}}{ad(a^2 + d^2) + 2b(a^3 + d^3)}\right\} \qquad ...(2.98b)$$

Expressions for losses in higher order modes of the resonator can be derived in a similar manner. In general Q increases when one goes for higher order modes.

2.3.3 Circular waveguide resonators

Analysis for circular waveguide resonators is very similar to that for rectangular waveguide resonators described above. Let us consider TE modes first. Magnetic field component H_z for the TE_{nm} mode in a cylindrical waveguide is given by

$$H_z = C_n J_n(h\rho)\cos n\phi\, e^{-j\beta z}$$

In the resonator, wave travelling in $-z$ direction is also to be considered and H_z may be written as

$$H_z = [C_n e^{-j\beta z} + C_n' e^{j\beta z}] J_n(h\rho)\cos n\phi \qquad ...(2.99)$$

If the length of the cavity is d, and we use boundary conditions at $z = 0$ and $z = d$, we can write H_z as

$$H_z = CJ_n(h\rho)\cos n\phi \sin\left(\frac{p\pi}{d}z\right) \qquad ...(2.100)$$

Resonant frequency can be obtained from

$$\frac{p\pi}{d} = \beta = \sqrt{\omega^2\mu\varepsilon - h_{nm}^2} \quad \text{or} \quad \frac{2\pi}{\lambda} = \left[h_{nm}^2 + \left(\frac{p\pi}{d}\right)^2\right]^{1/2} \qquad ...(2.101)$$

where $h_{nm} = (ha)'_{nm}/a$ and $(ha)'_{nm}$ are roots of $J'_n(ha) = 0$. Resonant condition for TM modes is obtained to be the same as (2.101) with h_{nm} given by $h_{nm} = (ha)_{nm}/a$, where $(ha)_{nm}$ are now roots of $J_n(ha) = 0$

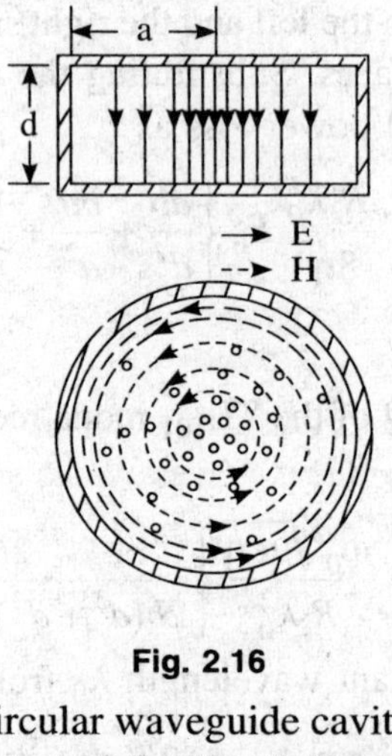

Fig. 2.16

The simplest mode in a circular waveguide cavity is TM_{010} mode. The non-zero field components for this mode are only E_z and H_ϕ and may be expressed as

$$E_z = CJ_0(h\rho)$$

$$H_\phi = \frac{j\omega\varepsilon}{h} J_1(h\rho) \qquad ...(2.102)$$

Since $p = 0$, resonance condition is given as

$$\sqrt{\omega^2\mu\varepsilon - h_{01}^2} = 0 \quad \text{or} \quad h_{01} = 2\pi/\lambda$$

Value of $(h_{02}a)$ is 2.405 (*see* eqn. 2.78). Resonant wavelength is obtained as

$$\lambda = 2.61a \qquad ...(2.103)$$

Field configuration for this mode is shown in Fig. 2.16. For this mode $\beta = 0$ and the dimension d can be made arbitrarily small. These resonators are known as cut-off resonators.

As we will subsequently come across in this text, the resonators of the type discussed above find applications in the design of several circuits like microwave oscillators and filters, etc.

SELECT BIBLIOGRAPHY

Discussions on topics contained in this chapter can be found in many books. Some of these are:

R.B. Adler, L.J. Chu and R.M. Fano, *Electromagnetic Energy Transmission and Radiation,* John Wiley and Sons, 1960. Chapter 3 on Steady-State Waves on Lossless Lines.

S Ramo, J.R. Whinnery and T. Van Duzer, *Fields and Waves in Communication Electronics,* John Wiley and Sons, 1965. Chapter 8 on Characteristics of Common Waveguides and Transmission Lines and Chapter 10 on Resonant Cavities.

E.C. Jordan and K.G. Balmain, *Electromagnetic Waves and Radiating Systems,* Prentice Hall Inc., 1968. Chapter 7 on Guided Waves and Chapter 8 on Wave Guides.

J.D. Kraus, *Electromagnetics,* McGraw-Hill Book Company Inc., 1953. Chapter 11 on Transmission Lines and Chapter 12 on Wave Guides.

J.D Kraus and K.R. Carver, *Electromagnetics,* McGraw-Hill Book Company, 1973. Chapter 13 on Transmission Lines, Waveguides and Resonators.

J.C. Slater and N.H. Frank, *Electromagnetism,* McGraw-Hill Book Company, 1947. Chapter 11 on Waveguides and Cavity Resonators.

R. Plonsey and R.E. Collin, *Electromagnetic Fields,* McGraw-Hill Book Company, Inc., 1961. Chapter 10 on Plane Waves, Waveguides and Resonators.

M. Javod and P.M. Brown, *Field Analysis and Electromagnetics,* McGraw-Hill Book Company, 1963. Chapter 12 on Propagation of Guided Waves, Transmission Lines and Rectangular Waveguides.

E. Argence and T. Kahan, *Theory of Waveguides and Cavity Resonators,* Blackie and Son Limited (London) 1967. Part III on Electromagnetic Waveguides and Part IV on Electromagnetic Cavities.

PROBLEMS

2.1 An air-filled coaxial transmission line has outer and inner radii equal to 6.903 mm and 3.0 mm, respectively. Calculate the values of (a) inductance per unit length, (b) capacitance per unit length and (c) characteristic impedance of the line.

2.2 A resistance $R = 10\ Z_0$ terminates a lossless line of characteristic impedance Z_0. The frequency of operation is increased by 1 per cent. Find the complex input impedance of the line when (a) the line length is $\lambda/2$ to begin with and (b) when the line is 25λ long to begin with.

2.3 An open-wire lossless line has the characteristic impedance Z_0 equal to 200 ohms. When it is short circuited, the first voltage minimum is 20 cms from the short circuited end. The short circuit is then removed and replaced by a load. The first voltage minimum is found to have moved to a point 11 cms from the load, and the *VSWR* is measured to be 3.33. What is the impedance of the load? (Use Smith chart for calculations.)

2.4 The load connected to the line in Problem 2.3 is to be matched to the line using a short circuited (smaller than $\lambda/4$) single stub. Find the location and length of the stub.

2.5 For a lossless transmission line terminated with a load, we have the characteristic impedance Z_0 equal to 100 ohms. The value of *VSWR* measured at the input is equal to 4.0 and first voltage maximum is $\lambda/8$ away from load. What is the load impedance? (Use Smith chart.)

2.6 To match the load in Problem 2.5, a $\lambda/4$ section of different line with characteristic impedance Z_{01} less than Z_0 is to be inserted somewhere along the original line. Find (i) the minimum distance (in terms of λ) between the load and the matching section and (ii) the characteristic impedance Z_{01}. (Use Smith chart.)

2.7 The transmission line of Problems 2.5 and 2.6 is a coaxial line with inner diameter of the outer conductor equal to 1 cm and air as dielectric. Find the diameter of inner conductor. If the $\lambda/4$ section of impedance Z_{01} is obtained by sliding a metallic sleeve over the inner conductor, find the outer diameter of the sleeve. (Metallic sleeves of this type are called slugs. See Fig. 4.25 for a typical application of slugs.)

2.8 Derive the relation (2.41) for the input impedance of a $\lambda/8$ transmission line section when terminated in a complex load $Z_L(= R + jX)$ with $Z_0 = |Z_L|$.

2.9 A rectangular waveguide is to be designed for operation in the frequency range 7.5 to 10 GHz. Calculate the inside dimensions so that the following design criteria are satisfied: (i) There is only one mode of propagation, (ii) lowest usable frequency is 10 per cent above the cut-off and (iii) highest usable frequency is 5 per cent below the frequency where next higher mode can propagate.

2.10 A rectangular waveguide with inside dimensions 0.9 inch by 0.4 inch is used for operation at 10 GHz. Waveguide walls are made of copper (surface resistivity $R_s = 2.6 \times 10^{-7}\sqrt{f}$ ohms). Calculate the attenuation constant in nepers/metre and dBs/metre.

2.11 For the waveguide described in Problem 2.10, find out the frequency at which the attenuation is minimum for the dominant mode. What is the value of this attenuation constant?

2.12 In a circular waveguide, the cut-off frequency for the dominant mode (with the lowest cut-off frequency) is 5 GHz. What is the frequency range over which only this mode propagates?

2.13 Provide a detailed derivation of the expression for the attenuation constant α in case of the dominant mode of rectangular waveguides as given in eqn. (2.73d) (p. 24).

2.14 Distribution of E and H fields for TM_{01} mode in a circular waveguide is shown in Fig. 2.12 (p. 28). Suggest a possible probe arrangement for excitation of this mode.

2.15 An air-filled transmission line with a characteristic impedance of 50 ohms is shorted at one end. The other end is terminated in a lumped capacitor of 4 pF. Calculate the length of the transmission line for resonance at 2 GHz.

2.16 Find the smallest possible size of cubical cavity with resonance at 3 GHz. In which mode (or modes) does it resonate at this frequency?

2.17 Derive expression for Q-factor of a cubical cavity operating in the lowest order mode.

2.18 A circular cylindrical cavity resonates at 3 GHz in TM_{010} mode. The resonator is filled with a dielectric material with dielectric constant 2.56. Calculate the new resonance frequency.

2.19 The waveguide described in Prob. 2.10 is used for designing a rectangular waveguide resonator. What should be the length of the resonator (dimension 'd') for TE_{101} mode resonance at 10 GHz? What is the next higher mode of resonance and the corresponding resonant frequency?

2.20 Find the quality factor Q of the resonator discussed in Prob. 2.19 at 10 GHz when the oavity is made of brass ($R_s = 5.01 \times 10^{-7} \sqrt{f}$ ohms). Also calculate the improvement in the Q-factor when the cavity is plated with silver ($R_s = 2.52 \times 10^{-7} \sqrt{f}$ ohms). Assume the silver thickness to be considerably larger than the skin depth.

CHAPTER 3

Generation of Microwaves by Vacuum Tubes

Like electromagnetic waves at lower frequencies, microwaves can also be generated by using vacuum tubes or semiconductor devices. However, these devices are based on different principles, although in some cases (such as for transistors) frequency limit has been extended into the microwave range by the advancement of technology and innovations in design.

This chapter describes principles of operation of the three most important microwave tubes—klystrons, magnetrons and travelling wave tubes. Generation of microwaves by semiconductor devices is discussed in Chapter 4.

3.1 LIMITATIONS OF CONVENTIONAL TUBES

3.1.1 Stray reactances

Oscillator circuits using triodes and pentodes exhibit several basic limitations when used at microwave frequencies. These limitations may be discussed by considering the case of a triode oscillator shown in Fig. 3.1.

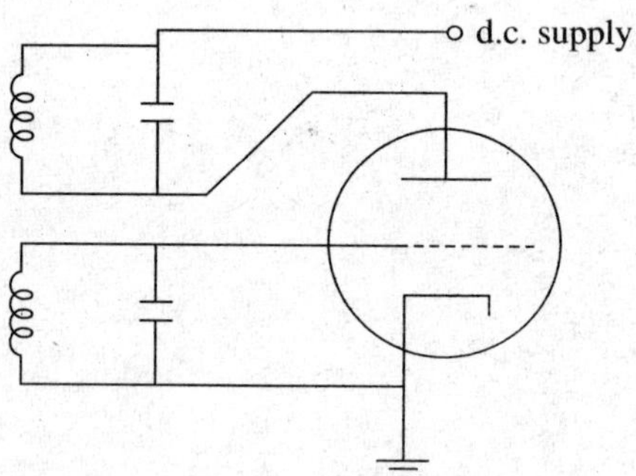

Fig. 3.1

As the operating frequency increases, values and hence sizes of inductors and capacitors need to be decreased. Values of these components become so small that the parasitic reactances of the tube and the leads used for connections become comparable. In fact, the tube and the lead reactances become so dominant around 500 MHz region, that it is not practical to employ any type of conventional capacitor or coil inductor. In this frequency region, sections of transmission lines (discussed in Chapter 2) are substituted for lumped element resonant circuits. Dual triode oscillators, such as shown in Fig. 3.2, use two-wire transmission lines (called lecher wires). Length l of the transmission line coupled to the plate and the plate-to-ground capacitance of the tube determines the resonant frequency. The feedback capacitors between the plates and the grids are not required because the tube has enough coupling within itself (plate to grid capacitances shown dotted in Fig. 3.2) to establish oscillations. These techniques allow one to develop higher frequency oscillators, but at about 1 GHz another factor namely transit time becomes significant and makes this approach also unsuccessful.

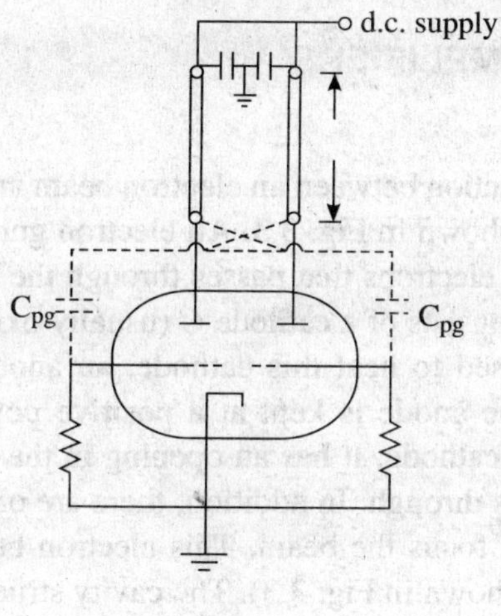

Fig. 3.2

3.1.2 Transit time

In a vacuum tube (like a triode) electrons emitted by the cathode take a finite (non-zero) time in reaching the anode. This interval, called the transit time, depends on the cathode-anode spacing and the static voltage between the anode and the cathode. Simple electron dynamics leads to

$$\tau = d / v_0 = d(V_e / 2m)^{1/2} \qquad ...(3.1)$$

where τ is the transit time, d is the cathode-anode spacing, v_0 is the velocity of electrons, V is the anode voltage, e is the electronic charge and m is the electronic mass. At lower frequencies transit time τ is a small fraction of the period of oscillation and is therefore insignificant. At microwave frequencies τ becomes comparable to the time period. This reduces the magnitude of the transadmittance and causes it to have an imaginary component. The output voltage and current are no longer in phase because of a phase-lag in the current produced by the finite transit time. This transit time phenomenon puts an upper frequency limit on the use of tubes like conventional triodes, etc.

Since the electron transit time decreases with the increase of the anode voltage and with the decrease of the inter-electrode spacing, the high frequency limit of operation can be extended by raising the operating voltage and/or by decreasing the inter-electrode spacing. Increase of the operating voltage, however, results in increased power dissipation at the anode, and decrease of the inter-electrode spacing increases the inter-electrode capacitance. If the inter-electrode spacing and electrode areas are decreased simultaneously, the allowable anode dissipation gets reduced. Thus the power output decreases with decrease of electrodes size required for high frequency operation.

However, in a suitably designed electron tube configuration (like klystron), the transit time phenomenon is employed usefully to generate microwave frequencies.

3.2 KLYSTRON AMPLIFIERS

3.2.1 Basic principle

Let us consider the interaction between an electron beam and an electromagnetic field of a cavity resonator as shown in Fig. 3.3. An electron gun is used to produce a high velocity narrow beam of electrons that passes through the fields in a resonant cavity.

The electron gun consists of a cathode *C* (usually oxide coated nickel cathode), a tungsten filament *F* used to heat this cathode, an anode and one or more beam focussing electrodes. The anode is kept at a positive potential and accelerates the electrons emitted by the cathode. It has an opening in the middle for the accelerated beam of electrons to pass through. In addition, there are one or more suitably shaped electrodes which help to focus the beam. This electron beam is later collected by a collector electrode (not shown in Fig. 3.3). The cavity structure shown in the figure is

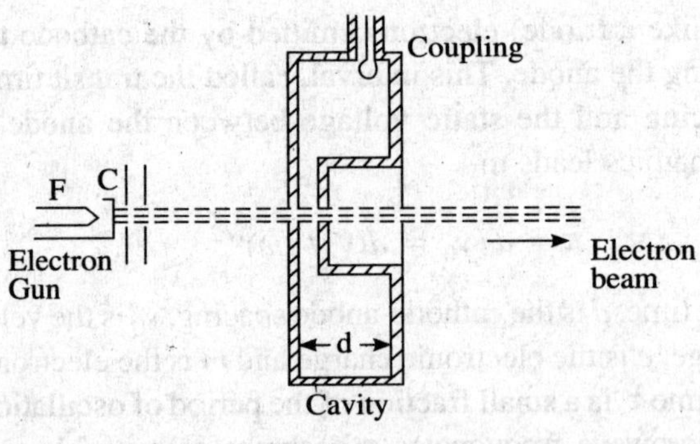

Fig. 3.3

a modification of the cylindrical cavity discussed in Chapter 2. The shape is symmetrical about the beam axis. The spacing between the top and bottom flat ends is reduced near the beam location. This helps to increase the magnitude of the electric field component at the beam location. This resonator structure is known as 're-entrant type cavity'. The electromagnetic power can be coupled into or out of the cavity through a coupling loop arrangement shown in the figure.

The electric field component in the cavity is collinear with the beam axis. Thus the electrons passing through the resonator see an accelerating field for half a cycle and a retarding field for the other half time period. Consequently, some of the electrons will be accelerated and some retarded. This phenomenon is called the velocity modulation of the electron beam.

As the electrons move ahead, at the average velocity of the beam, the velocity modulation causes a bunching of the electrons. The bunching effect converts the velocity modulation into what is known as the density modulation of the beam. This may be illustrated with the help of a distance-time diagram shown in Fig. 3.4. This representation is also known as Applegate diagram. Here each line represents distance-time history of an individual electron, the slope of each line being inversely proportional to the velocity of the electron it represents. Electrons entering the cavity at time instants *A, B* and *C* experience a zero value of the RF field and therefore travel without any change in velocity. Electrons that enter during the time interval *A* to *B* move slower and those that enter between *A* and *C* move faster. As a result of these differences in velocities, electrons tend to cluster around a point *D* at a distance *d.* The location of the point *D* depends upon the initial steady velocity of the electrons, RF field amplitude and the time period of the RF field. At the distance *d* the density of electrons in the stream varies periodically with time. Under these conditions, the electron beam contains an RF component of current and is said to be current modulated or density modulated. Thus we notice that as a result of the drift of the electron beam over a distance *d,* the velocity modulation has been converted into density modulation.

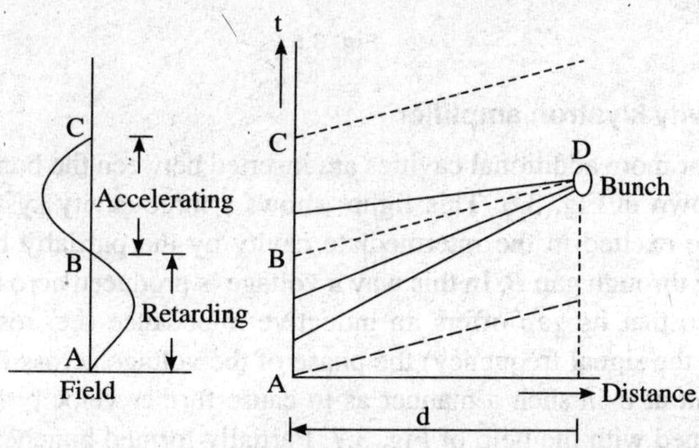

Fig. 3.4

If we place another resonant cavity at the distance d where the bunches are formed and an RF component of current is present, oscillations would be excited in the cavity and RF power can be brought out. This mechanism results in the conversion of DC kinetic energy of the beam into RF energy of the wave.

Klystron amplifiers are based on the principle discussed above. Essential components of a two-cavity klystron are shown in Fig. 3.5. This tube works as an amplifier with the input being fed to the buncher cavity and the output being obtained from the second cavity called the catcher cavity. It is also possible to connect the buncher and catcher cavities so that a small part of output power is fed back to supply the signal; in this way one can obtain an oscillator. Using the klystron as an oscillator in this manner would require a variable phase shifter between the two cavities to assure a positive feedback at the desired frequency of oscillation. This scheme is seldom used. As discussed later in Sec. 3.3, an oscillator can be designed using only a single resonant cavity and is known as reflex klystron oscillator. Reflex klystrons are very frequently used as microwave sources.

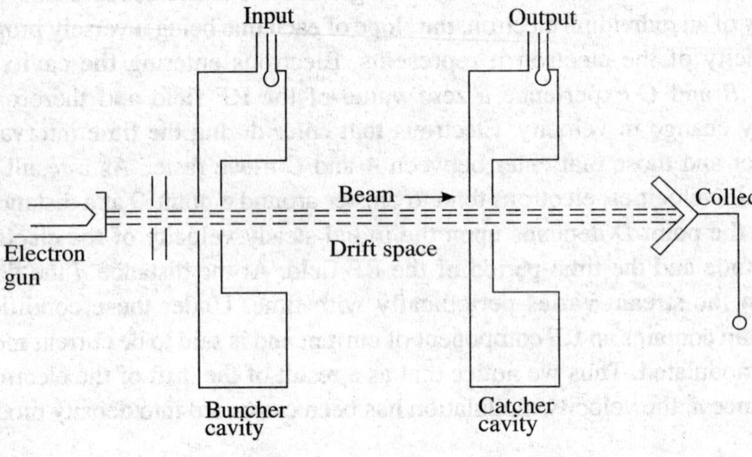

Fig. 3.5

3.2.2 Multicavity klystron amplifier

At times, one or more additional cavities are inserted between the buncher and catcher cavities as shown in Fig. 3.6. This figure shows a three-cavity system. In this case oscillations are excited in the intermediate cavity by the partially bunched electron stream passing through gap B. In this way a voltage is produced across B. By detuning the cavity B so that its gap offers an inductive impedance (i.e. resonant frequency slightly above the signal frequency) the phase of the voltage across B is related to the electron current at B in such a manner as to cause further velocity modulation. This may be explained with the help of Fig. 3.7. Partially formed bunches at time instants

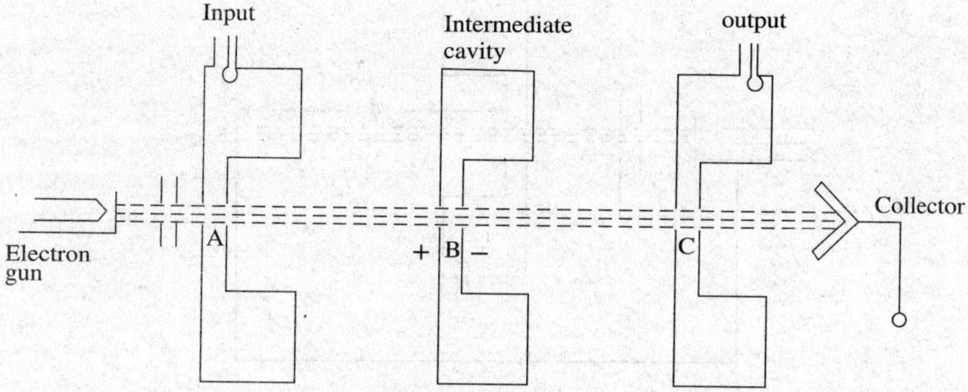

Fig. 3.6

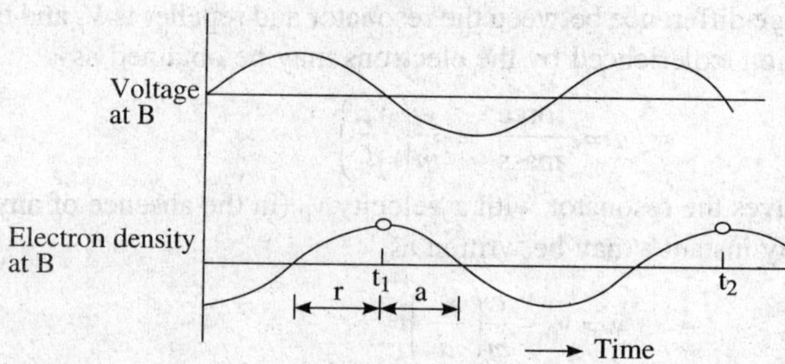

Fig. 3.7

t_1 and t_2 experience a retarding field in the time interval '*r*' ahead and an accelerating field in the interval '*a*' slightly later. This aids the bunching process and thus increases considerably the amplification available from the tube, and likewise raises the efficiency of the tube. It is also possible to increase the bandwidth of klystron amplifiers by employing one or more intermediate cavities that are appropriately stagger tuned.

Klystron amplifiers employing two or more cavities are available in various sizes ranging from units capable of delivering tens of milliwatts up to several hundred kilowatts of amplified output power.

3.3 REFLEX KLYSTRON OSCILLATOR

3.3.1 Single cavity configuration

A reflex klystron uses only a single cavity resonator and operates as an oscillator. Its basic configuration is shown schematically in Fig. 3.8. The various parts are : (i) Electron Gun, 1; (ii) Resonator, 2; (iii) Repeller, 3; and (iv) Output coupling, 4.

The repeller electrode is at a negative potential and sends the partially bunched electron beam back to the resonator cavity. This provides a positive feedback mechanism which supports oscillations.

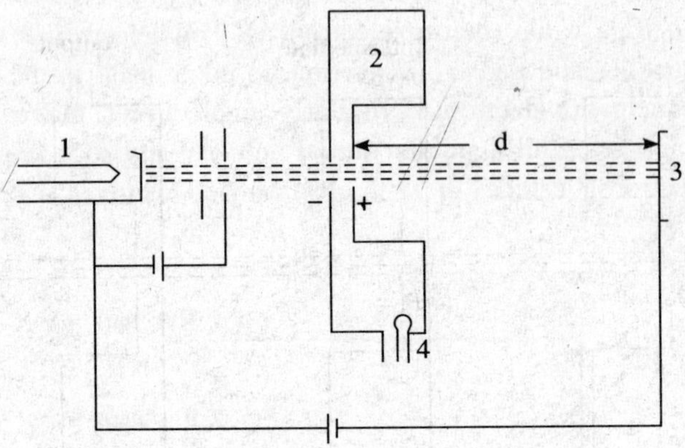

Fig. 3.8

If the voltage difference between the resonator and repeller is V_r and the distance is d, the retardation experienced by the electrons may be obtained as

$$a = \frac{\text{force}}{\text{mass}} = \frac{e}{m}\left(\frac{V_r}{d}\right) \quad \text{...(3.2)}$$

If an electron leaves the resonator with a velocity v_0 (in the absence of any RF field), its velocity at any instant t may be written as

$$v = v_0 - \frac{e}{m}\left(\frac{V_r}{d}\right)t \quad \text{...(3.2a)}$$

and its displacement x from the resonator will be

$$x = v_0 t - \frac{e}{m}\frac{V_r}{d}\frac{t^2}{2} \quad \text{...(3.3)}$$

This equation may be used to calculate the time t_0 taken by the electrons to return to the resonator. We obtain

$$t_0 = \frac{2mdv_0}{eV_r} \quad \text{...(3.4)}$$

It may be noted that t_0 is directly proportional to the initial steady velocity v_0.

3.3.2 Bunching in reflex klystrons

Bunching phenomenon in a reflex klystron can be visualized by studying electron trajectories in the region between the resonator and the repeller. Parabolic trajectories given by equation 3.3 are shown in Fig. 3.9. RF voltage across the resonator gap is also shown in this figure.

Let us consider an electron A which passes through the resonator gap when voltage across the gap is zero (changing from positive to negative). This electron returns to resonator at a time t_0 later. Another electron B passes through the gap Δt earlier. Since gap voltage is positive at this instant initial velocity v_0 for this electron is higher than the average value (that of electron *A*). Thus electron *B* would travel further toward the repeller and take a longer time to come back to the resonator. It would thus tend to bunch with electron *A*. Similarly, an electron *C* that passes through the gap Δt after *A*, appears with a smaller initial velocity and so takes less time to come back. Thus electron *C* catches up with electron *A*. This gives rise to a 'bunch' around the electron *A*.

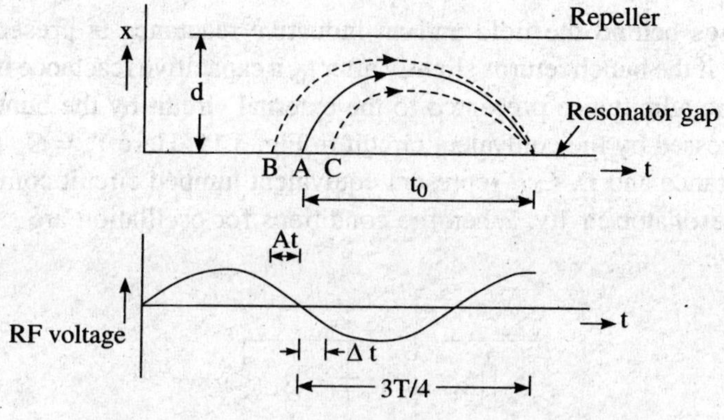

Fig. 3.9

The electron bunches formed as described above would deliver power to the resonator if they pass through the resonator at an instant when the field retards the bunch. Referring to the Fig. 3.9, we note that this happens when $t_0 = \frac{3}{4}\, T$(T being the time period at the resonance frequency). The above condition is, in general, satisfied for $t_0 = \left(n + \frac{3}{4}\right) T$, where *n* is any positive integer. Thus there are several values of t_0 satisfying the oscillation condition. These values correspond to various modes of a reflex klystron. A particular mode of operation may be selected by choice of the repeller voltage. Various modes of reflex klystron are presented in power output versus repeller voltage diagram (called the klystron mode curves) shown in Fig. 3.10. The

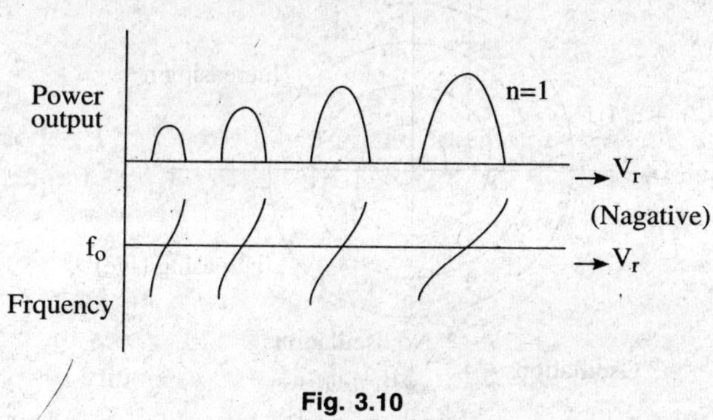

Fig. 3.10

mode corresponding to $n = 1$ occurs at maximum negative repeller voltage. For $n = 0$ (i.e. $t_0 = \frac{3}{4}\ T$), the gain mechanism is usually not strong enough to overcome system losses. Higher order modes $(n > 1)$ are present at lower voltages. Frequency versus repeller voltage variations shown in the figure are discussed in the following paragraph.

If the bunch returns to the cavity a little before the time given by $t_0 = \left(n + \frac{3}{4}\right)$ T, the current lags behind the field and an inductive reactance is presented to the circuit. Similarly if the bunch returns slightly after t_0, a capacitive reactance is presented to the circuit. The admittance presented to the external circuit by the bunches in the gap may be expressed by the equivalent circuit in Fig. 3.11. Here $Y_e = G_e + jB_e$ is the electronic admittance and C, L, R represent equivalent lumped circuit components of the microwave resonator cavity. When the conditions for oscillation are satisfied, G_e becomes negative.

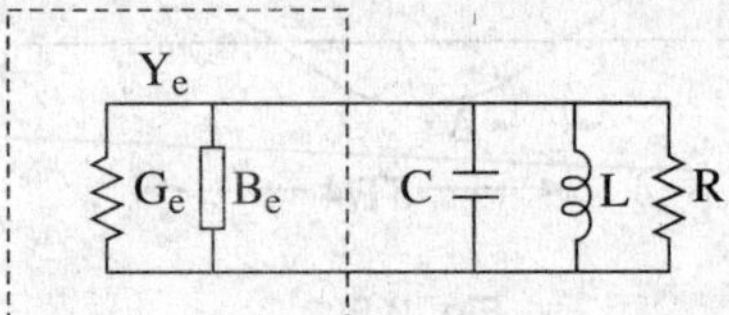

Fig. 3.11

Variation of G_e and B_e is shown qualitatively in Fig. 3.12, wherein Y_e is plotted in the complex admittance $(G_e + jB_e)$ plane. Infinite repeller voltage corresponds to the origin in this plane (zero admittance). As the repeller voltage is reduced one moves outwards along the spiral. Oscillations can take place whenever $| G_e |$ is greater than G_{ext}, the conductance of the external circuitry. If jB_e is negative, oscillations take place at a frequency higher than the resonance frequency of the cavity whereas positive values of jB_e correspond to a frequency lower than the cavity resonant frequency. This leads to the frequency variations with the repeller voltage as shown in Fig. 3.10.

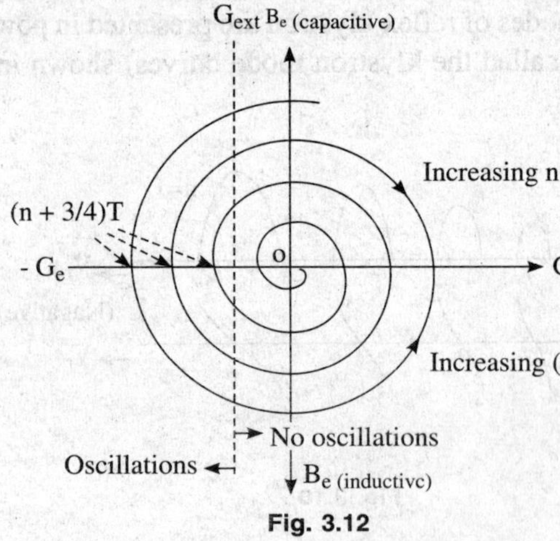

Fig. 3.12

It may be noted that after delivering the power to the resonant cavity, the electrons are collected by the body of the resonator which, for this reason, is kept at the most positive static voltage. Maximum power output of reflex klystrons is smaller than that available from two-cavity klystrons. Commercially available reflex klystrons range from small-size units producing 100 mW of power up to units capable of delivering several watts of power under continuous operation.

3.4 MAGNETRONS

3.4.1 Crossed electric and magnetic fields

Mechanism of the generation of microwaves in a magnetron involves interaction of electromagnetic fields with electrons moving in static electric and magnetic fields oriented at right angles to each other. Basic geometry and static fields are as shown in Fig. 3.13. Because of the positive voltage applied to the anode, the static electric field is in radial direction. Static magnetic field is in axial direction (right angle to E) and, in the present case, it is directed out of the plane of paper. When magnetic field is zero, the electron trajectory will be radial as shown by 'a' in the figure. When a small magnetic field B is applied the electron trajectory bends as shown by 'b'. The force exerted by the magnetic field is given by the relation $F = -e(\mathbf{v} \times \mathbf{B})$ The radius of curvature of the trajectory decreases as the magnetic field is increased. Above a certain value of the magnetic field, called the 'cut-off field', electrons are returned to the cathode and the anode collects no current. A typical electron path in such a case is indicated by 'd'. A magnetron is operated with magnetic field higher than the cut-off value.

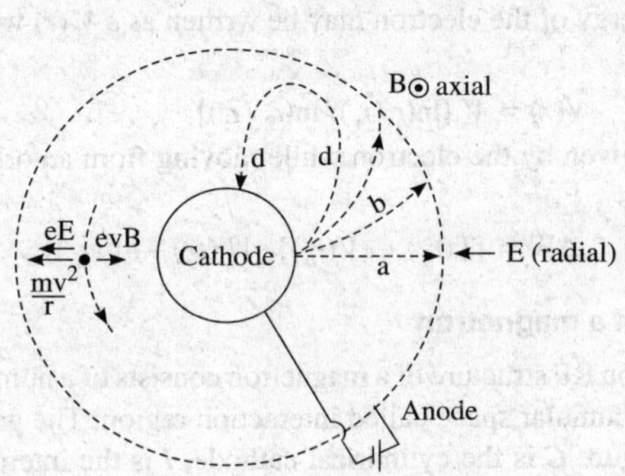

Fig. 3.13

There are three forces acting on an electron in the interaction region of the magnetron, force due to electric field ($-eE$), force due to magnetic field ($-e$ (v × B)) and the centrifugal force (mv^2/r). The resultant trajectory of the electron may be found by balancing these forces.

For discussing the mechanism of energy transfer in a magnetron, we onsider the case wherein the electrons move in a circular orbit. This is possible for a certain value of the initial velocity of the electrons. For the electron to move in a circular orbit of radius r (and concentric with the cathode anode geometry), the three forces mentioned above are radial as shown in Fig. 3.13, and balance when

$$mv^2/r + eE = evB \tag{3.5}$$

The electric field E is a function of r and of the applied voltage V_a. It may be expressed as

$$E = V_a/[r \ln(r_2/r_1)] \qquad \text{...(3.6)}$$

where r_2 and r_1 are anode and cathode radii, respectively. These two equations may be combined to obtain,

$$r = [(m/e)v + \{V_a/\ln(r_2/r_1)\}/v]/B \qquad \text{...(3.7a)}$$

The second term in (3.7a) is much higher than the first and consequently r increases when the electron velocity v is decreased. This characteristic of the electron motion would be useful in understanding the interaction mechanism taking place inside a magnetron.

A simplified model of energy transfer in crossed electric and magnetic fields can be obtained by writing electron radius r in terms of the kinetic energy $T(= mv^2/2)$. Equation 3.7a may be rewritten as

$$r = \frac{\sqrt{2m}}{eB}\left\{T^{\frac{1}{2}} + \frac{e}{2}T^{-\frac{1}{2}} \frac{V_a}{\ln(r_2/r_1)}\right\} \qquad \text{...(3.7b)}$$

Also, the potential energy of the electron may be written as $e\,V(r)$ where the potential $V(r)$ is given by

$$V(r) = V_a[\ln(r/r_1)/\ln(r_2/r_1)] \qquad \text{...(3.8a)}$$

Then the energy ΔW given by the electron while moving from an orbit of radius r_a to an orbit of radius r_b is

$$\Delta W = [T(r_a) + eV(r_a)] - [T(r_b) + eV(r_b)] \qquad \text{...(3.8b)}$$

3.4.2 RF structure of a magnetron

The most common RF structure of a magnetron consists of a number of cylindrical cavities opening in an annular space called interaction region. The geometry is shown in Fig 3.14. In this figure C is the cylindrical cathode, I is the interaction region and A is the anode block with multiple built-in cavities. The cavities are designed for TE modes with electric field lines in the plane of paper. The RF electric field configuration in a part of the structure is shown roughly in Fig. 3.15. The field distribution inside the cavity corresponds to TE_{110} mode of a cylindrical cavity. Even for similar field distributions in all the (eight) cavities shown, the magnetron structure supports varieties of modes depending upon the phase difference between fields in two adjacent cavities. Boundary conditions are satisfied when total phase shift around the eight cavities is a multiple of 2π radians. However, the most important mode for magnetron operation is one wherein the phase shift between the fields of adjacent cavities is π radians. This is known as π-mode. Frequency for π-mode can be easily separated from adjacent modes by incorporating conducting straps connected to alternate segments of the anode block. Field distribution for the π-mode in a portion of the interaction region near two adjacent cavities is shown roughly in Fig. 3.15. This is useful for discussing the mechanism of oscillations in magnetron. Several other RF structures are possible for designing a magnetron. However, the one shown in Fig. 3.14 is used very frequently.

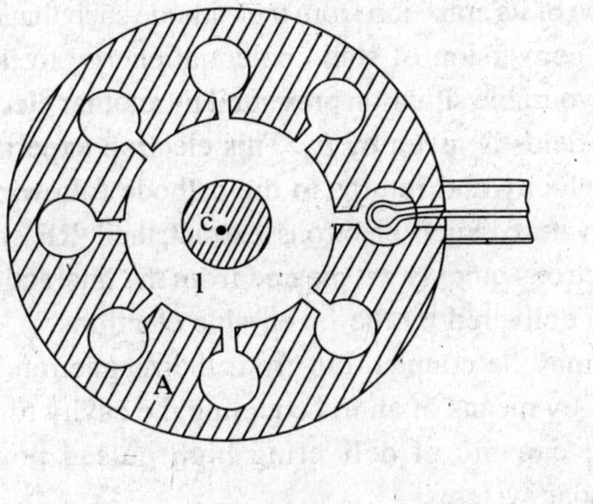

Fig. 3.14

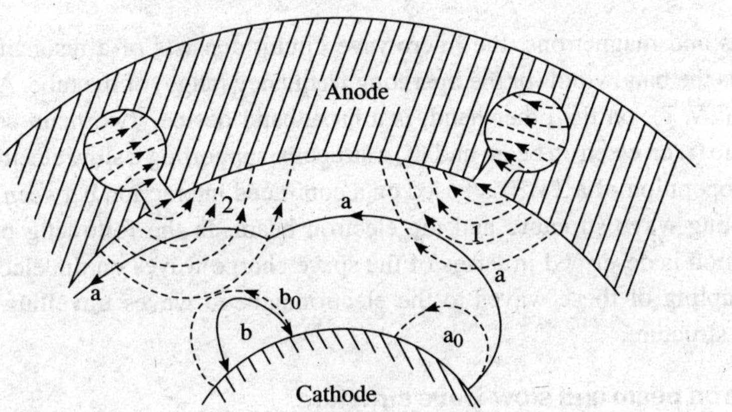

Fig. 3.15

3.4.3 Oscillation mechanism

Let us consider an electron emitted by the cathode and having a trajectory marked a_0 (in Fig. 3.15) in the absence of RF fields, and analyse the effect of RF fields of the π-mode on the motion of this electron. As this electron comes across an electric field in the direction of its velocity, it is retarded by the field, slows down and drifts towards the anode. Values of static E and H fields are so adjusted that by the time the electron reaches near the second cavity (marked 2), half a time-period elapses. The electron expriences a retarding field (opposite to that shown in the figure) again and loses energy to the RF field. This process continues and transfer of energy takes place again near the third cavity. Ultimately this electron strikes the anode surface after having delivered to the RF field a large part of its potential energy represented by its fall through the static potential difference between the anode and the cathode.

While discussing the case history of the electron described above we had selected its location and the time of its emission from the cathode such that the electron dynamics are favourable to the conversion of static potential energy to the RF energy. Let us now consider a less favourable situation presented by another electron whose trajectory in the absence of RF fields is given by b_0. This electron experiences an accelerating electric field, gains velocity and returns to the cathode following a trajectory b of a smaller radius of curvature. Such electrons extract their RF energy from the field. However, as these electrons quickly get cleared from the interaction region they absorb much less power than delivered by the favourable electrons.

The RF power may be coupled out from the magnetron by a coaxial loop as shown in Fig. 3.14, or by means of an iris coupling the cavity to a waveguide outside.

Magnetrons are capable of delivering high pulsed power outputs and are extensively used in radar systems.

3.5 TRAVELLING WAVE TUBES (TWT)

In klystrons and magnetrons, the microwave circuit consists of a resonant structure which limits the bandwidth (or the operating frequency range) of the tube. A travelling wave tube (TWT), on the other hand, is a broadband device. Two main constituents of a TWT are (i) an electron beam and (ii) a structure supporting a slow electromagnetic wave. The operation of a TWT is based on a continued interaction between the waves on a travelling wave structure and the electron beam. In the following paragraphs, this interaction is described in terms of the space charge waves on the electron beam and the coupling of these waves to the electromagnetic waves travelling along the slow-wave structure.

3.5.1 Electron beam and slow-wave structure

Electron beam in a TWT is generated by an electron gun similar to that used in klystrons. However, in a TWT, the electron beam is required to travel a much longer distance and a focussing mechanism becomes necessary to prevent the beam from spreading. The most common method of beam focussing is by applying a longitudinal static magnetic field along the axis of the beam. This magnetic field has no effect on the electrons travelling along or parallel to the beam axis. However, if an electron is diverging out and has a component of velocity in radial direction, the axial magnetic field produces a force on this electron. As can be seen by using Lorentz's force equation, the force due to this magnetic field is in the circumferential direction. Thus the electrons which try to diverge out due to space charge repulsion forces are bent into circumferential direction and start spiralling around the beam axis. This provides the required focussing mechanism.

The travelling wave structure used for TWT should have the two essential characteristics mentioned below: (i) the velocity of wave propagating along the structure should be smaller than the velocity of light so that the beam and the wave along the structure can travel with equal velocities and (ii) the wave should have an *E*-field component parallel to the beam axis so that interaction can take place between the electrons and the slow-wave longitudinal *E*-field.

These two requirements can be met by various structures. A helix (Fig. 3.16) is, however, most commonly used. The analysis of waveguiding properties of helix can be simplified by assuming it to be a 'sheath helix'. In a sheath helix the spacing between

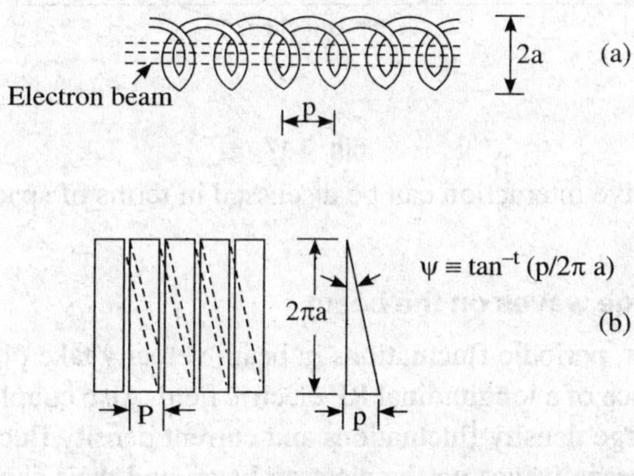

Fig.3.16

the adjacent turns, and the wire thickness are made to approach zero and the resultant structure becomes electrically smooth. This smooth 'sheath helix' is characterized by specifying that the conductivity in the direction of helix wire is infinite and that in the direction perpendicular to the helix wire is zero. The use of these boundary conditions allows a relatively simpler solution for the waves guided by the helix. We will not discuss the derivation of propagation constant based on this approach but simply point out the main results. It is found that the helix supports a slow wave with a phase velocity $v_p = c \sin \psi$ where c is the velocity of electromagnetic waves in free space and ψ is the helix angle given by $\tan^{-1} (p/2 \pi a)$. The helix has a radius 'a' and a pitch 'p'. The wave may be considered to propagate along the helical conductor with a velocity c. Thus the wave progresses along the axial direction z with a phase velocity $c \tan \psi$. The sheath helix model is valid at low frequencies, where the pitch p is much smaller than the wavelength.

When an electromagnetic wave is travelling along the helix, there is a longitudinal component of electric field. This time-varying field causes a velocity modulation in an otherwise uniform velocity of the electrons in the electron beam passing through the helix axis as shown in Fig. 3.16. The velocity modulation tends to bunch the electrons at regular intervals of one wavelength as shown in Fig. 3.17. When the wave and the electrons are moving with equal velocities, the position of electric field vis-a-vis the bunches in the beam remains invariant with time. Thus, a continuous interaction between the beam and the waves on the helix takes place and the bunches grow as the beam moves further.

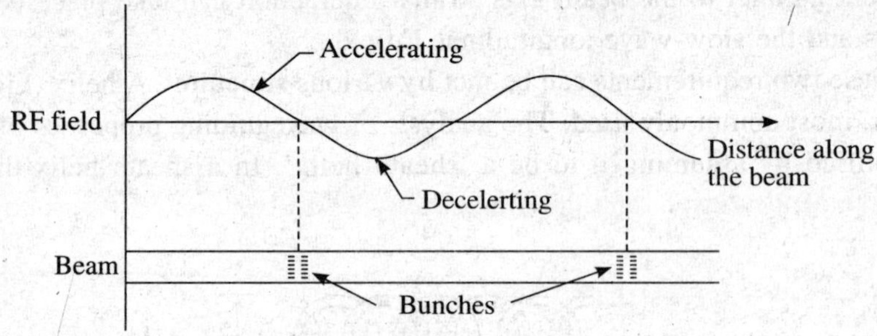

Fig. 3.17

This cumulative interaction can be discussed in terms of space charge waves on the beam.

3.5.2 Space charge waves on the beam

As indicated earlier, periodic fluctuations in beam velocity take place when the beam is under the influence of a longitudinal RF electric field. Also coupled to these velocity fluctuation are charge density fluctuations and current density fluctuations. These are known as space charge waves on the electron beam and their characteristics may be studied by a small signal analysis based on the laws of electron motion, the countinuity equation and the generalized Gauss's law.

Under small signal approximations, the various time-varying quantities may be expressed as

$$
\begin{aligned}
v &= v_0 + v_1\, e^{j(\omega t - \beta z)} \\
\rho &= \rho_0 + \rho_1\, e^{j(\omega t - \beta z)} \\
j_z &= J_{z_0} + J_{z_1}\, e^{j(\omega t - \beta z)} \\
E_z &= E_{z_1}\, e^{j(\omega t - \beta z)}
\end{aligned}
\qquad \text{...(3.9)}
$$

where β is the propagation constant for the wave and v_1, ρ_1 and j_{z1} are RF velocity, charge density and current density, respectively and are much smaller than corresponding static quantities v_0, ρ_0 and j_{z0}. Since current density J_z is the product of charge and velocity, we have

$$J_z = \rho v = \rho_0 v_0 + (\rho_1 v_0 + \rho_0 v_1) e^{j(\omega t - \beta z)} + \rho_1 v_1 e^{2j(\omega t - \beta z)} \quad ...(3.10)$$

The last term in this equation is much smaller than others and is neglected. This yields (since $J_{z0} = \rho_0 v_0$),

$$J_{z1} = \rho_1 v_0 + v_1 \rho_0 \quad ...(3.11)$$

Since electron motion is caused by E_z, we have

$$\frac{e}{m} E_z = \frac{dv}{dt} = \frac{\partial v}{\partial z}\frac{dz}{dt} + \frac{\partial v}{\partial t}$$

Carrying out the differentiations, one obtains

$$\frac{e}{m} E_{z_1} = j v_1 j_0 \left(\frac{\omega}{v_0} - \beta \right) \quad ...(3.12)$$

Continuity equation ($\nabla \cdot J_z = -\partial\rho/\partial t$) yields a relationship between the space derivative of the current and the time derivative of the charge which may be written as

$$j\beta J_{z1} = j\omega\rho_1 \quad ...(3.13)$$

Eliminating ρ_1 from equations (3.11) and (3.13) we get a relation for v_1 which when substituted in (3.12) yields a relationship between E_{z_1} and J_{z_1}. This gives

$$E_{z1} = j \frac{m v_0^2}{e \rho_0 \omega} \left(\beta - \frac{\omega}{v_0} \right)^2 J_{z1} \quad ...(3.14)$$

This relation may be written as

$$D_z = j \frac{v_0^2}{\omega_p^2 \omega} (\beta - \beta_0)^2 J_{z1} \quad ...(3.15)$$

where D_z is the Z-component of electric flux density D,

$$\beta_0 = \frac{\omega}{v_0} \quad ...(3.16)$$

and
$$\omega_p^2 = \frac{e\rho_0}{m\varepsilon_0} \quad ...(3.17)$$

$f_p (= \omega_p / 2\pi)$ is the plasma frequency. Its value is dependent on the electron density in the beam and may be written as

$$f_p = 8.98 \times 10^3 \, [n_e]^{1/2} \quad ...(3.18)$$

where n_e is the electron density in particles per cubic centimeter.

For low power TWTs f_p is of the order of 10^8 Hz. Time varying current density J_{z1} can be related to the radial component of D by applying divergence theorem to a small section of electron beam shown in Fig. 3.18. Applying the relation

$$\iint \left(J + \frac{\partial D}{\partial t} \right) da = 0 \quad ...(3.19a)$$

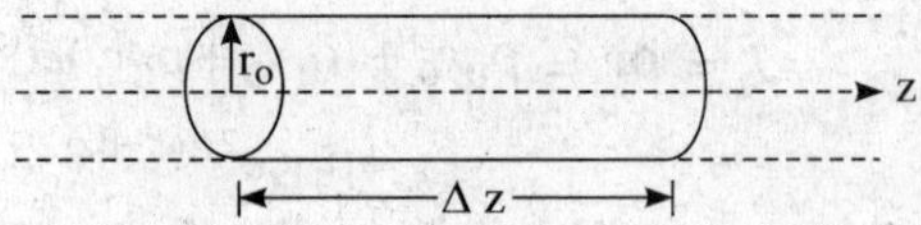

Fig. 3.18

to the closed area covering the volume shown in Fig. 3.18, we get

$$\left(\frac{\partial J_z}{\partial z} + j\omega\frac{\partial D_z}{\partial z}\right)\Delta z\pi r_0^2 + j\omega D_r 2\pi r_0 \Delta z = 0 \qquad \text{...(3.19b)}$$

Using (3.15) for D_z we have

$$J_{z1} = \frac{-2\omega\beta_p^2}{\beta r_0}\frac{D_r}{(\beta-\beta_0)^2-\beta_p^2} \qquad \text{...(3.20)}$$

where $$\beta_p = \frac{\omega_p}{v_0} \qquad \text{...(3.21)}$$

Since in general, D_r remains finite, there is a singular value of β at which the current becomes large.

Thus an effective current resonance occurs with the propagation constant given by

$$\beta = \frac{\omega \pm \omega_p}{v_0} \qquad \text{...(3.22)}$$

Equation 3.22 describes the propagation constants of the two space charge waves that may exist on the beam irrespective of the presence of external excitation. These represent normal modes of electron behaviour in the beam. These waves, called the slow and the fast space charge waves, travel at two different velocities given by

$$v = \frac{\omega}{\omega \pm \omega_p} v_0 \qquad \text{...(3.23)}$$

In the next section we will consider the interaction of these waves with the electromagnetic wave travelling along the slow wave circuit.

3.5.3 Coupling between beam and slow-wave circuit

The slow-wave structure may be represented by a transmission line model shown in Fig. 3.19. We can write

$$j\beta_1 = (YZ)^{\frac{1}{2}} \qquad \text{...(3.24)}$$

where Y and Z represent the shunt admittance and the series impedance per unit length for the slow-wave structure. Phase constant for the circuit is given by β_1. The coupling between the beam and the structure is considered by incorporating a displacement current di as shown in Fig. 3.19. In the presence of this coupling we have

$$dV = -IZdz$$

and $$dI = -VYdz - di \qquad \text{...(3.25)}$$

where $di = (\partial i/\partial z)dz$ and i is the beam current given by $J_{z1}\pi r_0^2$. Replacing d/dz by $-j\beta$, these lead to a relationship between β (phase constants for coupled waves) and β_1 (given by eqn. 3.24) as

$$(\beta^2 - \beta_1^2)V = j\beta Zi = j\beta Zj_{z1}\pi r_0^2 \qquad ...(3.26)$$

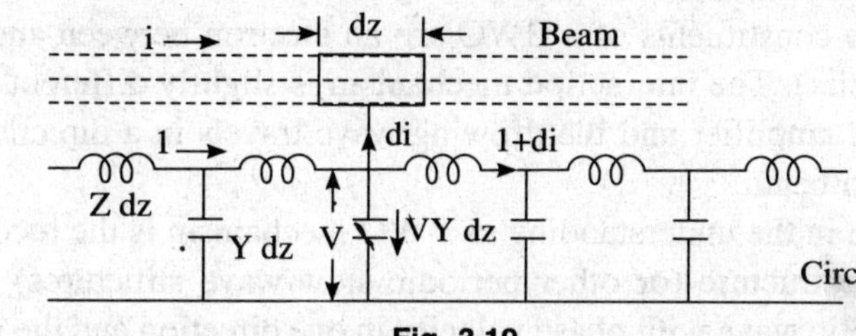

Fig. 3.19

Voltage V is proportional to transverse component of electric field which is expressed in terms of J_{z1} by (3.20). Substituting for J_{z1} from (3.20) leads to

$$(\beta^2 - \beta_1^2) = \frac{-A}{(\beta - \beta_0)^2 - \beta_p^2} \qquad ...(3.27)$$

where A is a constant given by

$$A = \frac{-2\omega\beta_p^2 X\varepsilon_0 E_r}{r_0 V}\pi r_0^2 \qquad ...(3.28)$$

Negative sign in (3.28) is introduced as an *ad hoc* assumption relating V and E_r, and we have used $Z = jX$. Equation (3.27) is of fourth order in β and therefore, has four roots. For operation of a TWT, β is nearly equal to β_1 (for the circuit) and also β_0(for the beam). Since $\beta_p << \beta_0$, we can see from (3.27) that there are three forward wave solutions and one backward wave solution corresponding to the factor $(\beta + \beta_1)$. If we replace $(\beta + \beta_1)$ by $3\beta_1$ (3.27) may be written as

$$(\beta - \beta_1)(\beta^2 - 2\beta\beta_0 + \beta_0^2 - \beta_p^2) = -(A/2\beta_1) \qquad ...(3.29)$$

Let us examine a particular case when $\beta_1^2 = \beta_0^2 - \beta_p^2$. We may then write (since $\beta_0 \cong \beta_1$):

$$(\beta - \beta_1)^3 = -(A/2\beta_1) \qquad ...(3.30)$$

This leads to three values of β as

$$\begin{aligned} \beta &= \beta_1 + b(\cos\pi/3 + j\sin\pi/3) \\ \beta &= \beta_1 + b(\cos\pi + j\sin\pi) \qquad ...(3.31) \\ \beta &= \beta_1 + b(\cos 5\pi/3 + j\sin 5\pi/3) \end{aligned}$$

where $b = (A/2\beta_1)^{1/3}$. Three values correspond to three types of variation along z: (*a*) $e^{-j(\beta_1+0.5b)z}e^{-0.866bz}$ which represents a wave that is attenuated exponentially as it travels, (b) $e^{-j(\beta_1-b)z}$ which represents an unattenuated wave and, (c) $e^{-j(\beta_1+0.5b)z}$ $e^{+0.866bz}$ which represents a wave that grows exponentially as it travels. This wave (c) dominates the other two for large values of z and is responsible for amplification in a TWT.

TWT amplifiers are known for their wide bandwidth and low noise characteristics. They find extensive applications in microwave communication systems.

3.5.4 Backward wave oscillators

A backward-wave oscillator (BWO) is a self-oscillating TWT capable of delivering microwave power over a wide frequency range. As in the case of a travelling wave amplifier, the main constituents of a BWO are an electron between and a slow-wave structure (e.g. a helix). The interaction mechanism is slightly different from that in a conventional TWT amplifier and the growing wave travels in a direction opposite to that of the electron beam.

The first step in the understanding of BWO mechanism is the recognition of the fact that the helix structure (or other periodic slow-wave structures) can support a slow electromagnetic wave with phase velocity in one direction and the group velocity (or power flow) in the opposite direction. A typical dispersion curve ($\omega - \beta$ diagram) for a helix structure is as given in Fig. 3.20. Other periodic slow-wave structures also show similar behaviour. In the portion *AB* of the dispersion curve both the phase

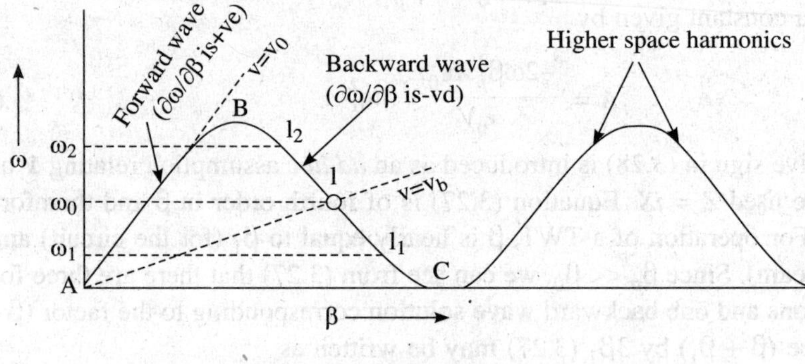

Fig. 3.20

velocity (ω/β) and the group velocity ($\partial\omega/\partial\beta$) are positive. This corresponds to the forward wave used for interaction in the conventional TWT amplifier. As shown in the figure an electron beam with $v = v_a$ will be synchronous with this wave. However, in the portion *BC* of the dispersion curve, we note that phase velocity is positive but the group velocity ($\partial\omega/\partial\beta$) is negative. This behaviour characterizes a backward wave which is useful for interaction in a backward wave oscillator. An electron beam with a velocity v_b will be synchronous with the backward wave at point *I* corresponding to a frequency ω_b. It may be noted that this frequency of interaction can be varied from (say for example) ω_1 to ω_2 when the electron beam velocity is varied such that the interaction point shifts from I_1 to I_2. This property makes BWO a voltage tunable oscillator.

The operation of the BWO can be visualised in terms of a series of feedback loops similar to those common in low-frequency electronics circuits. This arrangement is shown in Fig. 3.21. Each of these regenerative loops can function as an amplifier or an oscillator and is designed so that the phaseshift around the loop is one cycle. Using conventional terminology, the forward or μ-circuit consists of a section of transmission line and the backward or β-circuit is a unilateral amplifier connecting the output of the transmission line to the input. Positive feedback takes place in this circuit when the amplifier gain becomes sufficiently high to overcome the loss in the transmission line, and the μβ loop oscillates at a frequency for which the total phase delay is one cycle. In a backward wave oscillator the unilateral amplification is provided by the wave along the electron beam and μ-circuit corresponds to the backward wave on the helix. Starting with an electric field at a point on the helix near the electron gun, this field velocity modulates the electron beam, which moves forward toward the collector and forms a bunch. This bunch supplies energy to the helix wave that is carrying its power in the gun direction. Thus the amplified wave gets back to the original point, completing the feedback loop. When there is sufficient current in the electron beam, this feedback mechanism produces oscillations at the frequency where the electron velocity is in step with the backward wave of the helix.

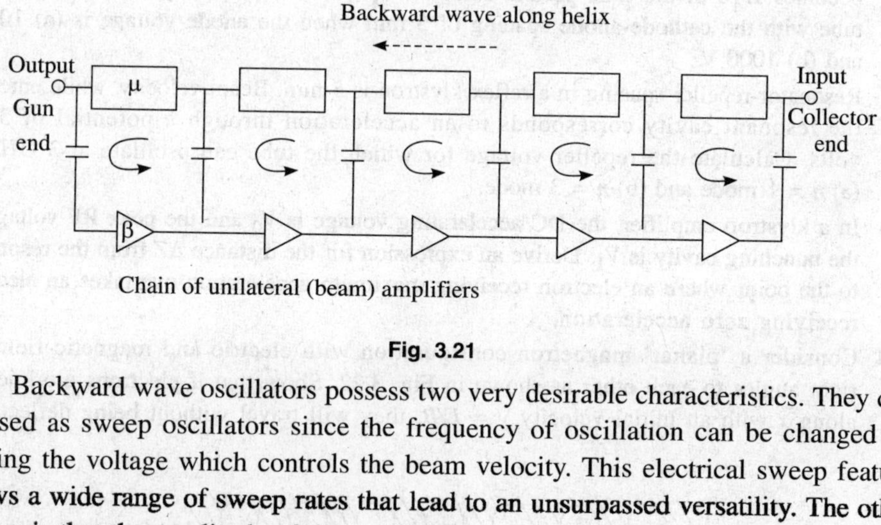

Fig. 3.21

Backward wave oscillators possess two very desirable characteristics. They can be used as sweep oscillators since the frequency of oscillation can be changed by varying the voltage which controls the beam velocity. This electrical sweep feature allows a wide range of sweep rates that lead to an unsurpassed versatility. The other feature is that the amplitude of the oscillation can be decreased continuously to zero by changing the beam current. This feature is unusual in oscillators and can be readily used to provide amplitude modulation or to provide a constant output signal by incorporating a feedback control.

A backward wave tube can also be used as a voltage-tunable band pass amplifier (with input at the collector end and output at the helix end) when operated with beam currents below oscillation threshold levels.

SELECT BIBLIOGRAPHY

R.E. Collin, *Foundations for Microwave Engineering,* McGraw-Hill Book Company, 1966. Chapter 9 on Microwave Tubes.

H.A. Atwater, *Introduction to Microwave Theory,* McGraw-Hill Book Company, 1962. Chapter 6 and 7 on Microwave Tubes.

H.J. Reich, *Microwave Principles,* Van Nostrand Reinhold Company, 1957 and Affiliated East-West Press Pvt. Ltd., New Delhi, 1972. Chapter 10, Klystrons; Chapter 11, Magnetrons and Chapter 12, Travelling wave and electronwave tubes.

Proceedings I.E.E.E., Vol. 61, No. 3, March, 1973. (This issue contains four invited papers on high power microwave tubes.)

G.D. Sims and I.M. Stephenson, *Microwave Tubes and Semiconductor Devices* Inter- science Publishers New York, 1963.

PROBLEMS

3.1 If useful frequency range of a triode is limited to a point where the transit time becomes 1/10 of the time period, calculate up to what frequency can one operate a tube with the cathode-anode spacing of 5 mm when the anode voltage is (a) 100 V and (b) 1000 V.

3.2 Resonator-repeller spacing in a reflex klystron is 5 mm. Beam velocity, while entering the resonant cavity corresponds to an acceleration through a potential of 3000 volts. Calculate the repeller voltage for which the tube can oscillate at 3 GHz in (a) $n = 1$ mode and (b) $n = 3$ mode.

3.3 In a klystron amplifier, the DC accelerating voltage is V_0 and the peak RF voltage at the bunching cavity is V_1. Derive an expression for the distance ΔZ from the resonator to the point where an electron receiving maximum acceleration overtakes an electron receiving zero acceleration.

3.4 Consider a 'planar' magnetron configuration with electric and magnetic fields at right angles to each other as shown in Fig. 3.22. Show that if electrons are injected along x with an initial velocity $v = E/B$, they will travel without being deflected.

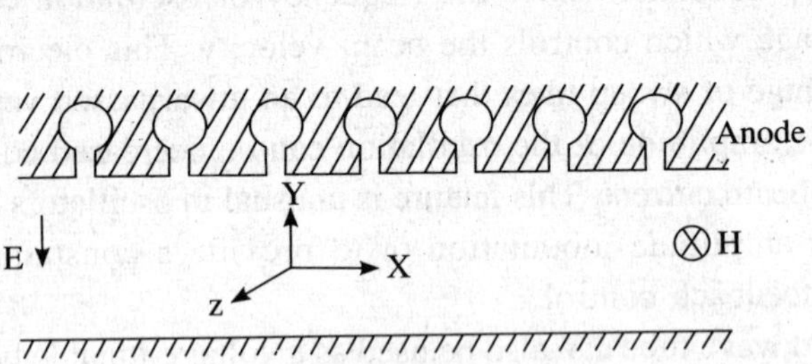

Fig. 3.22

3.5 Consider the configuration shown in Fig. 3.22. Assume that the electrons start from cathode with zero initial velocity. Electric field is 250 V/mm. If the cathode-anode spacing is 2 mm find the value of critical magnetic field above which the electrons are not collected by the anode. (*Hint:* Equations for electron motion may be written as

$$\frac{d^2y}{dt^2} = \left(a - \omega_c \frac{dx}{dt}\right) \text{ and } \frac{d^2x}{dt^2} = \omega_c \frac{dy}{dt}$$

where $a = -eE/m$ and $\omega_c = Be/m$.)

3.6 Show that an electron with velocity v perpendicular to B_0 (no electric field present) executes circular motion at the cyclotron frequency $\omega_c = eB_0/m$. (**Hint** : Equate centrifugal force to $-e\mathbf{v} \times \mathbf{B_0}$ force).

3.7 A magnetron with cathode and anode diameters of 1 and 1.2 cm, respectively, operates at 3 GHz in magnetic flux density of 1600 gauss. If there are 8 anode segments and the device operates in the π-mode, what are the values of the beam velocity and the anode voltage. Magnetron can be approximated by the planar configuration shown in Fig. 3.22 and electron position considered mid-way of cathode and anode.

3.8 Consider an electron beam with an average velocity 2×10^7 cm/sec. The electron beam is excited by an electric field at 3 GHz. Calculate the current density for which the fast space charge wave along the beam travels at a velocity higher than that of the beam by 10 per cent. What will be the velocity of the slow-space charge wave?

3.9 In a helix-type travelling wave tube operating at 5×10^8 Hz, the pitch to the circumference ratio of the helix is 1:10, and the beam density is such that the plasma frequency is $\omega_p = 2\pi \times 10^8$ sec^{-1}. What accelerating voltage is required for the electron gun in order that the condition $\beta^2_1 = \beta^2_0 - \beta^2_p$ shall be fulfilled?

3.10 Growth constant for a wave along an active TWT is 3 dB/cm. If the current density of the beam is made double (all other parameters remaining unchanged), how will the growth constant be modified?

CHAPTER 4

Microwave Solid State Sources

Similar to the difficulties experienced in the design of vacuum tubes (triodes etc.) for microwave frequency range, there have been problems in extending the operating frequency of transistors and other semiconductor devices. Here also, the transit time of carriers through the device becomes significant and has to be reduced. Also, the capacitances associated with the junctions become large and have to be brought down. However, during the last decade, these difficulties have been overcome and the transistors available today are capable of delivering respectable powers at microwave frequencies. At 8 GHz a silicon *n-p-n* transistor with 3.9 dB noise figure and 3.8 dB gain, and a power transistor with 1 Watt output power and 6 dB power gain have been reported. Microwave field effect transistors, using GaAs and a metal-semiconductor Schottky junction for the gate contact, have also made considerable advances in recent years. Currently these transistors (called GaAs-MESFETs) can deliver a 4 Watt output at 4 GHz with a 6 dB power gain and a 44% power-added efficiency. At 22 GHz available MESFETs can provide a 140 mW power output with a 4.8 dB power gain and 9% power-added efficiency.*

In addition to transistors semiconductor devices based on newer principles are available for generating power at higher microwave frequencies. These are two terminal devices and there are two distinct families. The first category consists of transferred electron devices (TEDs) that employ differential negative mobility *(dv/dE)* of electrons in certain *n*-type III-V compounds, particularly gallium-arsenide (GaAs), for their operation. These include Gunn diodes and LSA diodes. These are bulk devices in which the interaction occurs in the bulk of single type of semi-conductor material. Best reported CW power of transferred electron oscillators (TEO) range from 1.5 Watts at X-band (8 to 12 GHz) and 0.5 Watts at Ku-band (12 to 18 GHz) to about 20 mW at 88 GHz. The pulsed power output that can be obtained from TEOs operating in LSA mode are impressive. These devices have been reported to deliver 6 KW at 1.75 GHz with 15% efficiency,

---

* Data regarding capabilities of transistors and MESFETs are extracted from reference [11].

2 KW at 7 GHz with 5% efficiency and 150 Watts at 16 GHz with 6% efficiency.*

The second type of microwave oscillator diodes are avalanche diodes. These are also negative resistance devices but their negative resistance results from mechanisms considerably different from those in TEDs. These are junction devices and the negative resistance is obtained by appropriately combining the impact avalanche breakdown process and the carrier transit-time effect. The highest CW power output of these devices ranges from about 8 Watts at 8 to 10 GHz to about 100 mW at 94 GHz. These power outputs are significantly higher than the best CW results of TEDs. Also, avalanche devices can operate at considerably higher frequencies than TEDs. Another mode of operation of avalanche devices called TRAPATT mode is capable of delivering high pulsed power outputs. State of the art figures are 240 Watts for 0.2 μs pulses, 0.1% duty cycle, 8.3% 1 dB bandwidth and 29% efficiency in frequency range 3 to 4 GHz.**

This chapter discusses the basic principles of these microwave semi-conductor devices.

4.1 BIPOLAR TRANSISTORS

4.1.1 General considerations

Basic operation of microwave bipolar transistors is identical to that for transistors at low frequencies. Microwave power transistor circuits may be discussed by considering the power amplifier block diagram shown in Fig. 4.1. As shown in this figure, the input and the output matching is obtained by using impedence transformers. Design of various types of impedance transformers will be treated later in Chapter 7. For the present discussions, these networks are assumed to be lossless. The input power is given by

$$P_{\text{in}} = i_{in}^2 R_{in} \qquad \text{...(4.1)}$$

and the output power may be written as

$$P_{\text{out}} = i_{out}^2 R_{out} \qquad \text{...(4.2)}$$

Power gain is obtained to be

$$P_0/P_{\text{in}} = [i_{out} / i_{in}]^2 [R_{out} / R_{in}] \qquad \text{...(4.3)}$$

Ratio $i_{\text{out}}/i_{\text{in}}$ is the AC current gain factor denoted by h_{fe}. It is also expressed as f_t/f, where f_t is the cut-off frequency corresponding to a current gain value of unity, and f is the operating frequency. Equation 4.3 may be rewritten as

$$\text{Power gain} = (f_t / f)^2 (R_{out} / R_{in}) \qquad \text{...(4.4)}$$

---

* From B. Berson, 'Active microwave solid state device performance', presented at 1975 IEEE-MTT-S International Microwave Symposium. Palo Alto (Calif.), June 1975.

** From E.D. Cohen, 'TRAPATTS and IMPATTS, State of the art and applications', Microwave J., Vol. 20, Feb. 1977, pp. 22-28.

Thus the transistor parameters that influence power gain are: (i) f_t (should be large), (ii) R_{out} (should be large) and (iii) R_{in} (should be small). We shall discuss the dependence of these parameters on transistor design in a qualitative manner.

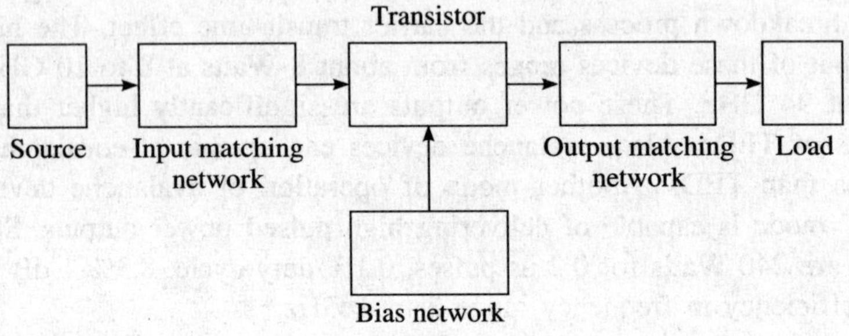

Fig. 4.1

Vertical cross-section of a typical microwave transistor is shown in Fig. 4.2. Transistor fabrication is carried out by a set of processes known as 'planar technology'. These processes are identical to those used for low frequency transistors. A wafer of n^+ silicon is used as the substrate. A thin layer of *n*-type silicon is grown over this substrate by 'epitaxial growth'. This constitutes the *n*-region for the collector. An insulating layer of SiO_2 is thermally grown over the *n*-region. The base-layer diffusion pattern is then cut into the silicon dioxide layer by employing the photo-resist method. The base layer is formed by diffusion of *p*-type impurities. After the *p*-type base diffusion, the emitter-layer pattern is formed using similar methods. The emitter junction can be formed by diffusion (in which case the device is called a double-diffused transistor) or by alloying (in which case the device is called a diffused-base alloyed-emitter transistor). A final metallization process is used to make ohmic contacts to the emitter and the base. The collector contact is from the bottom surface of the n^+ substrate.

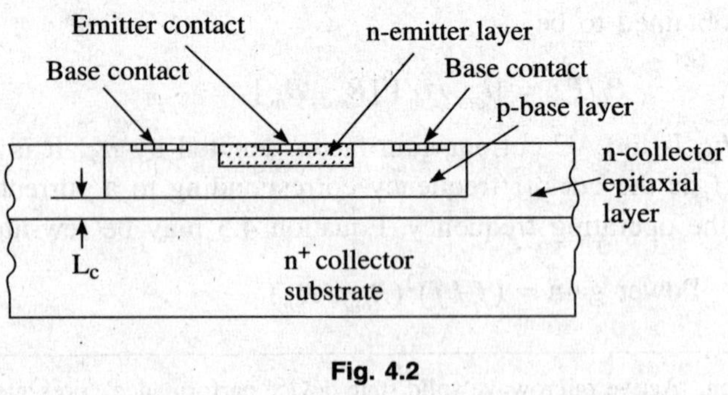

Fig. 4.2

4.1.2 Cut-off frequency (f_t)

Like vacuum tubes, transistors also have their high frequency operation limited by the transit time. In vacuum tubes, the transit time is the time taken by electron emitted by cathode to reach anode. In the case of bipolar transistors transit time is the time taken by the charge carriers (electrons for ***npn*** and holes for ***pnp*** transistors) to move from the emitter to the collector. Based on the transit time of the charge carriers, a cut-off frequency called transit time cut-off frequency (f_t) may be defined as

$$f_t = 1/(2\pi\tau) \qquad ...(4.5)$$

where τ is the transit time.

This time of transit from emitter to collector is composed of four parts: (i) Emitter charging time τ_e, (ii) base transit time τ_b, (iii) collector space charge transit time τ_{sc} and (iv) collector charging time τ_c.

Various parameters that determine the above mentioned components of the transit time would now be discussed.

(*a*) *Emitter charging time:* It is the product of the resistance and transition capacitance of forward biased emitter-base junction

$$\tau_e = r_e C_e \qquad ...(4.6)$$

The resistance r_e is determined by the *I–V* characteristic of a *p-n* junction as follows. We have

$$I_e = I_0(e^{qV/kT} - 1) \qquad ...(4.7)$$

or

$$\frac{\partial I_e}{\partial V} = q\frac{I_e + I_0}{kT} \simeq \frac{qI_e}{kT} \qquad ...(4.8)$$

where q is the electronic charge (1.602×10^{-19} coulombs), k is the Boltzmann constant (1.38×10^{-23} joules/deg) and T is the temperature in degrees Kelvin. Thus,

$$r_e = \frac{kT}{qI_e} = \frac{26}{I_e} \text{ milliohms (at 300°K)} \qquad ...(4.9)$$

The capacitance C_e is the junction capacitance given by

$$C_e = C_0\left(1 - \frac{V}{V_1}\right)^{1/n} \qquad ...(4.10)$$

where V_i is the contact potential and n, a constant whose value is 2 for an abrupt junction and 3 for a linearly graded junction and C_0 is the junction capacitance at zero bias. In addition to C_e given by (4.10), stray capacitance (caused by the package, etc.) and Miller capacitance given by $C_c(1 + g_m R_L)$ and to C_e.

Emitter charging time forms a considerable fraction of the total transit time. τ_e can be reduced by decreasing the emitter area and increasing the operating current I_e. A value of τ_e near 10 picosecond is typical for an f_t of 6 GHz.

*(b) **Base transit time:*** When the base is uniformly doped, the transit of carriers through the base is solely by diffusion. If D is the diffusion constant and W the base width, the base transit time may be written as

$$\tau_b = W^2/2D \qquad \text{...(4.11)}$$

This expression can be derived from the definition of the diffusion constant D. Diffusion current is related to the diffusion constant as

$$J = -Dq\frac{\partial n}{\partial x} \qquad \text{...(4.12)}$$

If the doping in the base region is not uniform there is a built-in electric field. This electric field arises because of non-uniform distribution of acceptors in (p-type) or donors (in n-type) and may be evaluated by using Gauss's law. In this case, the base transit time τ_b may be written as

$$\tau_b = \frac{W^2}{\eta D} \qquad \text{...(4.13)}$$

where η is a function of the built-in field. For a diffused-base alloyed-emitter structure typical value of η is 4. Typical value of τ_b for a 6 GHz cut-off transistor is 2 picosecond.

(c) Collector space charge transit time (τ_{sc}) : At the operating bias, the depletion layer in the collector region has a finite width which is determined by the impurity concentration in the collector body and the reverse voltage applied to the collector-base junction. At the high fields existing in the depletion layer, the velocity of carriers traversing the region saturates at a value determined by ionized impurity scattering. If the depletion width is X and the saturation velocity of carriers, v_s, the transit time is given by X/v_s. This assumes that electrons get accelerated to the velocity v_s over a distance much smaller as compared to X. One factor has not been considered yet in the simple formulation. The AC voltage across the depletion region excites a displacement current in this region. This causes a phase shift in the total current flowing across the depletion layer. Consequently, effective transit time is reduced.

Typical value of τ_{sc} for a cut-off frequency of 6 GHz is 12 picosecond.

(d) Collector charging time τ_c : Collector charging time constant is also an RC product, namely r_cC_i where r_c is the resistance in series with the collector depletion layer capacitance C_i. In a conventional diffused transistor, where the collector body has a resistivity determined by the doping level necessary to achieve the desired breakdown voltage, r_c is determined by the spreading resistance and is usually so high that r_cC_i product is one of the major limiting factors in frequency response.

One may minimize r_c by the use of an epitaxial structure as shown in Fig. 4.2. In this configuration the collector depletion layer extends completely across the high resistivity epitaxial layer. In the case of such a swept-out collector region $r_c = \rho_c L_c / A$, where ρ_c is the resistivity of the substrate and is kept very low. L_c

is its thickness and L_c is the area. In such a case typical value of τ_c for a 6 GHz cut-off frequency is 1 ps.

To sum up this discussion, it may be pointed out that the transit time may be decreased by (a) decreasing the emitter area and thus reducing the emitter charging time, (b) having non-uniform doping in base region so that the built-in field reduces the base transit time, and (c) using an epitaxial collector for reducing the collector series resistance.

4.1.3 Output resistance (R_{out}) and input resistance (R_{in})

As seen in equation 4.4 resistances R_{out} and R_{in} are the other two important parameters. Output resistance R_{out} is given by the load line slope. The load line is a function of the output power, the class of operation, and the operating voltage and, as such, is independent of the transistor. However, it is shunted by the collector capacitance. In order to have a large output impedance the collector capacitance must be minimized. This can be achieved by making the base area (and hence the effective collector area) smaller.

The input resistance R_{in} is governed by the emitter geometry and the spacing between emitter and base contacts. At microwave frequencies skin effect causes most of the emitter current to flow near its edges and therefore, R_{in} is inversely proportional to the emitter periphery. Also R_{in} is directly proportional to the distance between the emitter and the base contact.

Thus, in order to obtain an optimum performance from a microwave transistor, emitter periphery should be maximized and the base area should be minimized. Consequently, an important parameter in transistor performance is the ratio of emitter periphery to base area. This ratio is commonly known as aspect ratio.

4.1.4 Geometries for microwave transistors

There are a number of horizontal geometries suited for microwave transistors. The most important of these are: (a) Interdigitated, (b) Overlay, (c) Emitter mesh and (d) Diamond emitter. The maximum aspect ratio obtainable from these geometries depends upon the minimum width that can be defined by photolithography for purposes of diffusion, etc.

Interdigitated geometry shown in Fig. 4.3 has been widely used for small signal and power transistors. Consider a unit cell of this geometry consisting of half of an emitter stripe, one space (between the adjacent fingers) and half of a base stripe (shown dotted in the inset in Fig. 4.3). If the length of the stripes is ten times the width W, base area of unit cell is $10W \times 2W$ or $20W^2$; the emitter periphery for the same area is $11W$. This gives an aspect ratio of $11(20W)$.

Overlay and mesh geometries are used for power transistors only. An overlay configuration is shown in Fig. 4.4. In this design, the metal base fingers (of structure in Fig. 4.3) are replaced with a highly conductive p^+ diffusion. The p^+ stripes are connected together with a wide metallic base contact. Emitter stripes

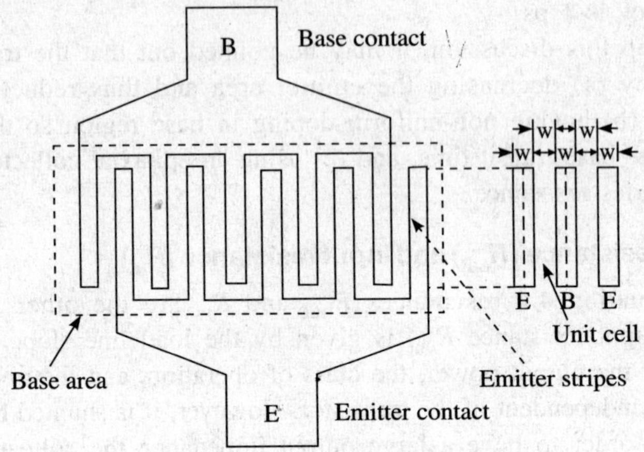

Fig. 4.3

are also seen in this figure. Emitter metallization runs across the base stripes also, but makes contact only with the emitter stripes as the base stripes are separated by an insulating (SiO_2) layer. The geometry does not provide any improvement in aspect ratio when compared with the interdigitated structure. An emitter mesh configuration is a modification of the overlay structure shown in Fig. 4.4. In place of emitter stripes a mesh of square shaped emitters is used.

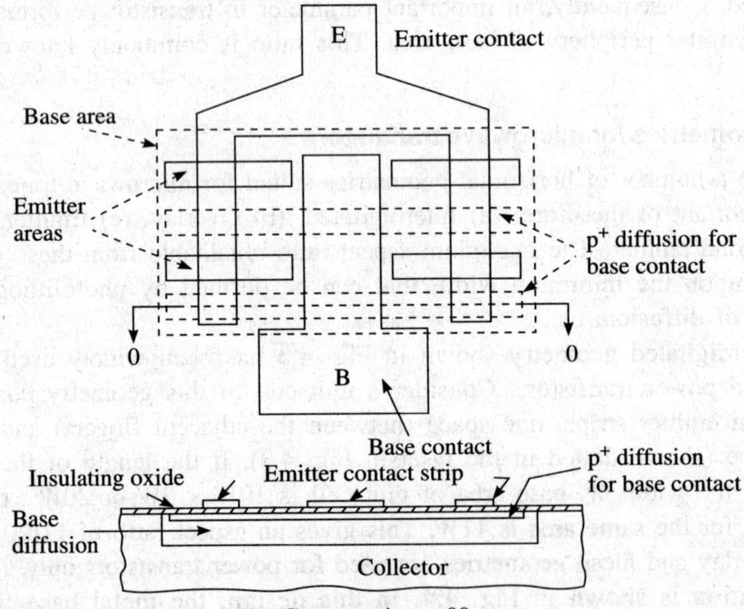

Fig. 4.4

Diamond emitter configuration (Fig. 4.5b) is a modification of the 'square' overlay structure shown in Fig. 4.5a. It consists of a checker board pattern with

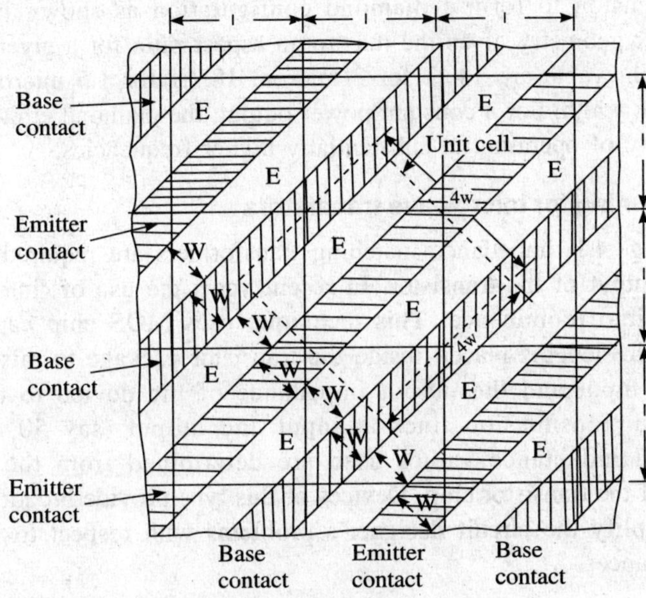

Fig. 4.5(a)

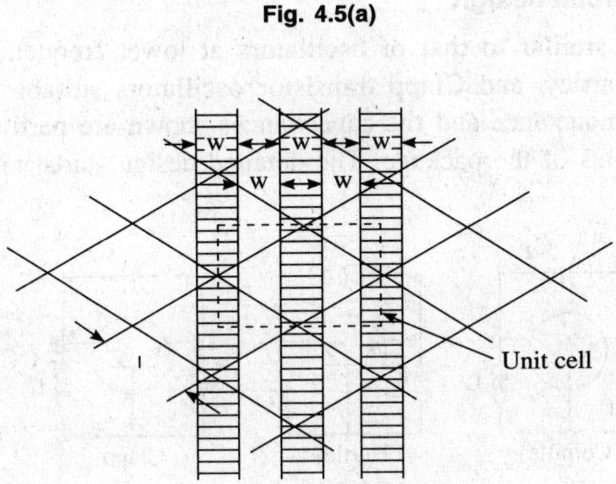

Fig. 4.5(b)

alternate squares representing the emitter areas. These emitter areas are marked with letters E in Fig. 4.5a. A unit cell of the structure is shown enclosed in the dotted square of $4W \times 4W$. The emitter also has a square shape, the dimensions of which are $l \times l$ or $\sqrt{8}(W) \times \sqrt{8}W$. Metallization stripes of width W run diagonally and provide emitter and base contacts alternately. As shown previously in Fig. 4.4 an insulating layer of SiO_2 prevents emitter contact stripes to make contact with base areas and vice versa. Calculation of aspect ratio is straightforward. Base area, for the unit cell, is 16 W^2 (area of the unit cell itself) and the emitter periphery is given by $4\sqrt{8}W$. This gives an aspect ratio of $1/\sqrt{2}W$. Longer emitter periphery can be obtained if the squares of the checker board pattern are elongated along one dimension to form a diamond configuration as shown in Fig. 4.5b. Diamond emitter geometry gives the maximum aspect ratio for a given line width. Aspect ratios achievable are 14 :1 for 2 micron, 18.7:1 for 1.5 micron and 28 :1 for 1 micron line width. For a constant power output, the diamond emitter transistor has the capability of operating at substantially higher frequencies.

4.1.5 Internal tuning for microwave transistors

As shown in Fig. 4.1, impedance matching transformers are required at both the input and the output of the transistor. In recent years the use of 'internal tuning' has also come into prominence. This technique uses MOS chip capacitors and discrete bond wire lengths placed inside the transistor package to raise either one or both of the input and the output impedance of the device to the level of commonly used transmission lines at input and output (say 50 ohms). The capacitance and inductance values used are determined from the impedance characteristics of the transistor chip. Devices of this type provide broader bandwidth and greatly simplify the circuit designer's problems with respect to matching to very low impedances.

4.1.6 Oscillator circuit design

Basic designs are similar to that of oscillators at lower frequencies. Figure 4.6 shows Colpitts, Hartley, and Clapp transistor oscillators suitable for microwave applications. The inductance and the capacitances shown are partly contributed by the parasitic elements of the package. The detailed design starts with measurement

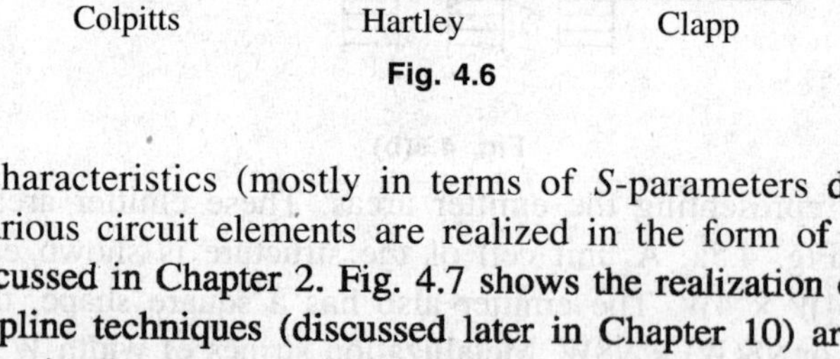

Fig. 4.6

of transistor characteristics (mostly in terms of S-parameters discussed later in Chapter 5). Various circuit elements are realized in the form of transmission line sections as discussed in Chapter 2. Fig. 4.7 shows the realization of Colpitts circuit at 2 GHz. Stripline techniques (discussed later in Chapter 10) are used. The base of the transistor is directly grounded to the ground plane of the stripline board. The capacitance C_1 is provided by the base to emitter capacitance of the device. Feedback is provided by the phase-resonant loop with L_2 and C_2. In the output

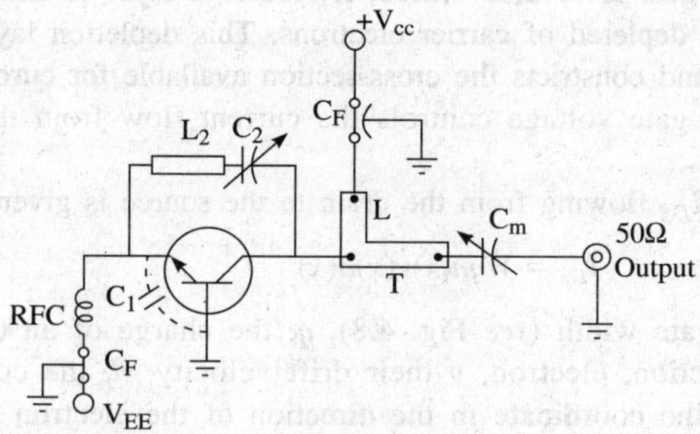

Fig. 4.7

circuit L provides the necessary reactance to tune the output capacitance and T is a quarter wave transformer which transforms the real part of the collector impedance to about 50 ohms. C_m provides adjustment of the matching network. C_F represents feed through capacitor for the DC bias.

4.2 FIELD EFFECT TRANSISTORS

4.2.1 MESFET configuration

Field effect transistors (FETs) at microwave frequencies are mostly fabricated in GaAs and use a metal-semiconductor (MES) Schottky junction for the gate contact. The basic configuration is shown in Fig. 4.8. It consists of a moderately doped n-type epitaxial layer on a high resistivity (semi-insulating) GaAs substrate. Two

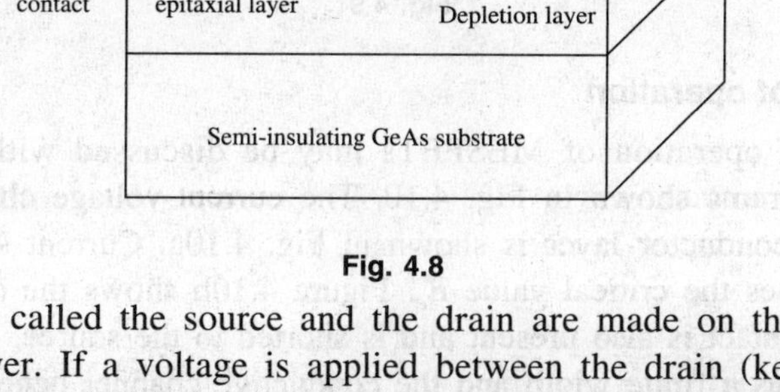

Fig. 4.8

ohmic contacts, called the source and the drain are made on the top surface of the epitaxial layer. If a voltage is applied between the drain (kept positive) and the source, electrons will flow from source to the drain. Hence, the source acts as the origin of carriers and drain as a sink. Another contact, a metal semiconductor Schottky junction, is added between the source and the drain. This is called the 'gate'. When the gate is reverse biased, it creates a layer in the semiconductor that is completely depleted of carrier electrons. This depletion layer acts like an insulating region and constricts the cross-section available for current flow in the n-layer. Thus the gate voltage controls the current flow from the drain to the source.

The current I_{DS} flowing from the drain to the source is given by

$$I_{DS} = Wqn(x)v(x)d(x) \qquad \text{...(4.14)}$$

where W is the gate width (*see* Fig. 4.8), q, the charge of an electron, n, the density of conduction, electron, v their drift velocity, d, the conductive layer thickness and, x the coordinate in the direction of the electron drift. The drift velocity v of the carriers is a function of electric field E in the semiconductor. Variations of drift velocities for GaAs and silicon are shown in Fig. 4.9. For silicon the drift velocity increases monotonically from zero value to a saturation v_s. For n-type GaAs the drift velocity of carriers exhibits a peak value v_p at E equal to E_p. For values of E greater than E_p, the slope is negative which results in a negative value of the differential mobility. Presence of this negative differential mobility makes the analysis of GaAs MESFET operation more complicated. Therefore, we will first discuss the MESFET operation for silicon devices.

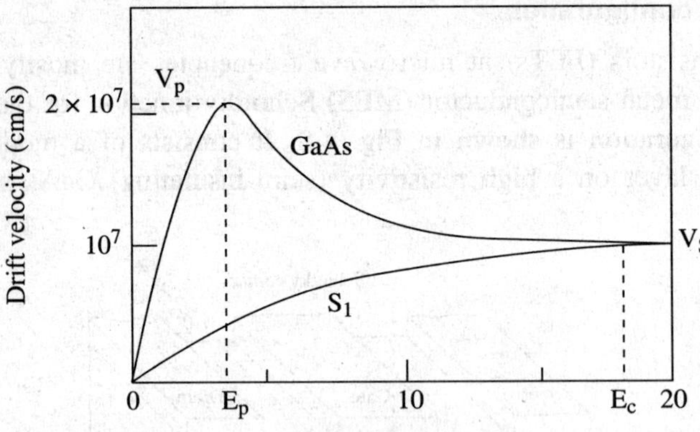

Fig. 4.9

4.2.2 Principle of operation

The principle of operation of MESFETs may be discussed with the help of a sequence of diagrams shown in Fig. 4.10. The current-voltage characteristic of a thin n-type semiconductor layer is shown in Fig. 4.10a. Current starts to saturate when E approaches the critical value E_c. Figure 4.10b shows the current variation when the gate contact is also present and is shorted to the source. In this case, the depletion layer has a finite width and the conductive channel beneath has a cross-section d smaller than in Fig. 4.10a. The voltage along the channel increases from zero at the source to V_{DS} at the drain. Thus the metal-to-semiconductor junction

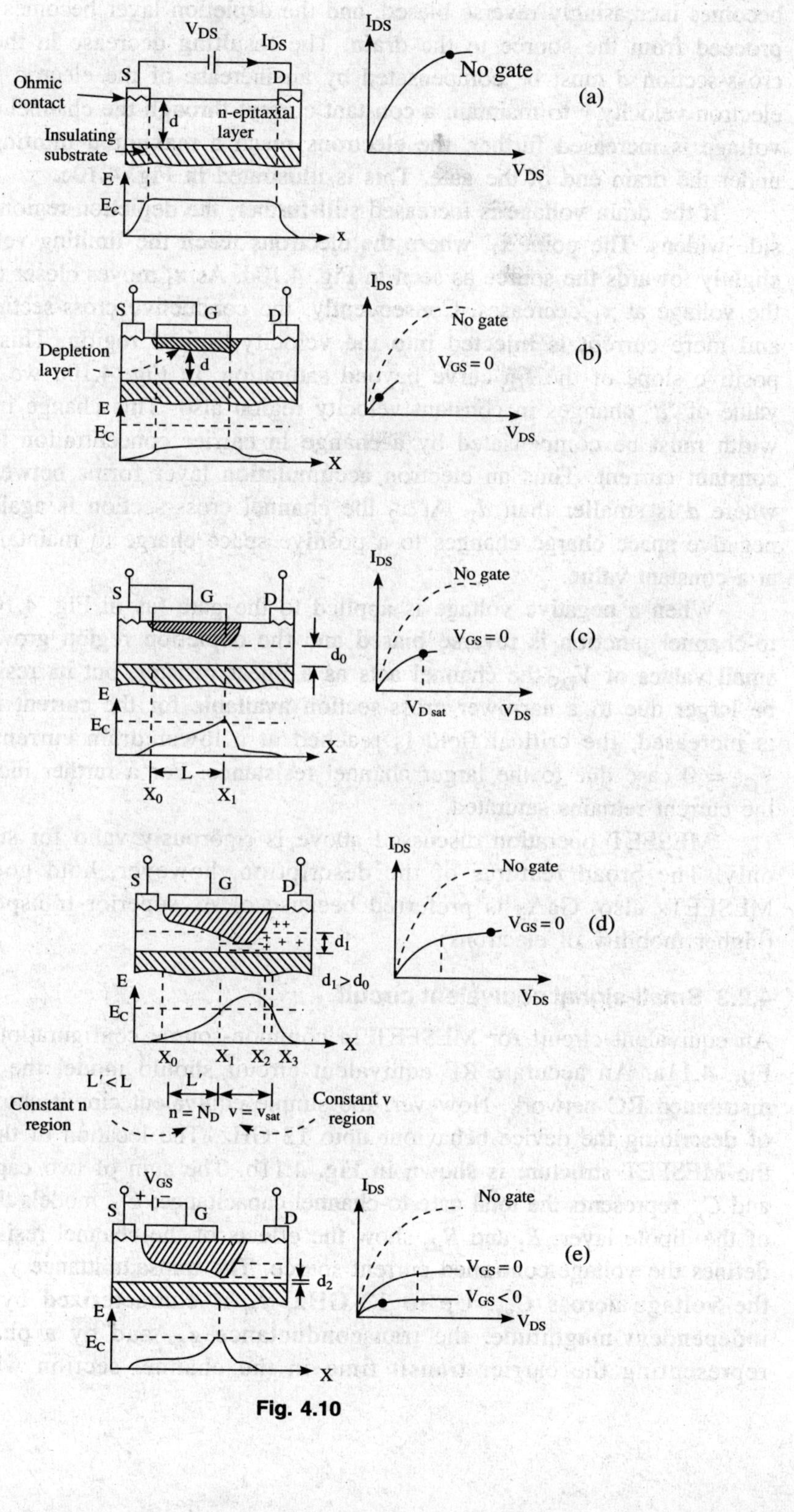
V_DS
I_DS
S
D
Ohmic contact
d
n-epitaxial layer
Insulating substrate
E
E_c
x
I_DS
No gate
(a)
V_DS
S
G
D
Depletion layer
d
E
E_C
X
I_DS
No gate
V_GS = 0
(b)
V_DS
S
G
D
d_0
E
E_C
X
L
X_0
X_1
I_DS
No gate
V_GS = 0
(c)
V_D sat
V_DS
S
G
D
++
+ + +
d_1
d_1 > d_0
E
E_C
X_0
X_1
X_2
X_3
X
L' < L
L'
Constant n region
n = N_D
v = v_sat
Constant v region
I_DS
No gate
V_GS = 0
(d)
V_DS
V_GS
+
S
G
D
d_2
E
E_C
X
I_DS
No gate
V_GS = 0
V_GS < 0
(e)
V_DS

Fig. 4.10

becomes increasingly reverse biased, and the depletion layer becomes wider as we proceed from the source to the drain. The resulting decrease in the conductive cross-section d must be compensated by an increase of the electric field and the electron velocity v to maintain a constant current through the channel. As the drain voltage is increased further, the electrons reach a maximum limiting velocity v_s under the drain end of the gate. This is illustrated in Fig. 4.10c.

If the drain voltage is increased still further, the depletion region on the drain side widens. The point x_1, where the electrons reach the limiting velocity, moves slightly towards the source as seen in Fig. 4.10d. As x_1 moves closer to the source, the voltage at x_1 decreases. Consequently, the conductive cross-section d_1 widens and more current is injected into the velocity-limited region. This results in a positive slope of the I_{DS} curve beyond saturation. In Fig. 4.10d we note that the value of 'd' changes in constant velocity region also. This change in the channel width must be compensated by a change in carrier concentration to maintain a constant current. Thus an electron accumulation layer forms between x_1 and x_2 where d is smaller than d_1. At x_2 the channel cross-section is again d_1 and the negative space charge changes to a positive space charge to maintain the current at a constant value.

When a negative voltage is applied to the gate (as in Fig. 4.10e), the gate-to-channel junction is reverse biased and the depletion region grows wider. For small values of V_{DS} the channel acts as a linear resistor, but its resistance would be larger due to a narrower cross-section available for the current flow. As V_{DS} is increased, the critical field is reached at a lower drain current than in the $V_{GS} = 0$ case due to the larger channel resistance. For a further increase in V_{DS}, the current remains saturated.

MESFET operation discussed above is rigorously valid for silicon devices only. The broad features of the description, however, hold good for GaAs MESFETs also. GaAs is preferred because of its superior transport properties (higher mobility of electrons).

4.2.3 Small-signal equivalent circuit

An equivalent circuit for MESFET in common-source configuration is shown in Fig. 4.11a. An accurate RF equivalent circuit should model the channel as a distributed RC network. However, the simple equivalent circuit shown is capable of describing the device behaviour upto 12 GHz. The location of the elements in the MESFET structure is shown in Fig. 4.11b. The sum of two capacitances C_{dg} and C_{gs} represents the total gate-to-channel capacitance; C_{dc} models the capacitance of the dipole layer; R_i and R_{ds} show the effects of the channel resistance; and i_{ds} defines the voltage-controlled current source. The transadmittance y_m relates i_{ds} to the voltage across C_{gs}. Up to 12 GHz, y_m is characterized by a frequency independent magnitude, the transconductance g_m, and by a phase delay τ_θ, representing the carrier transit time in the channel section where $E > E_c$.

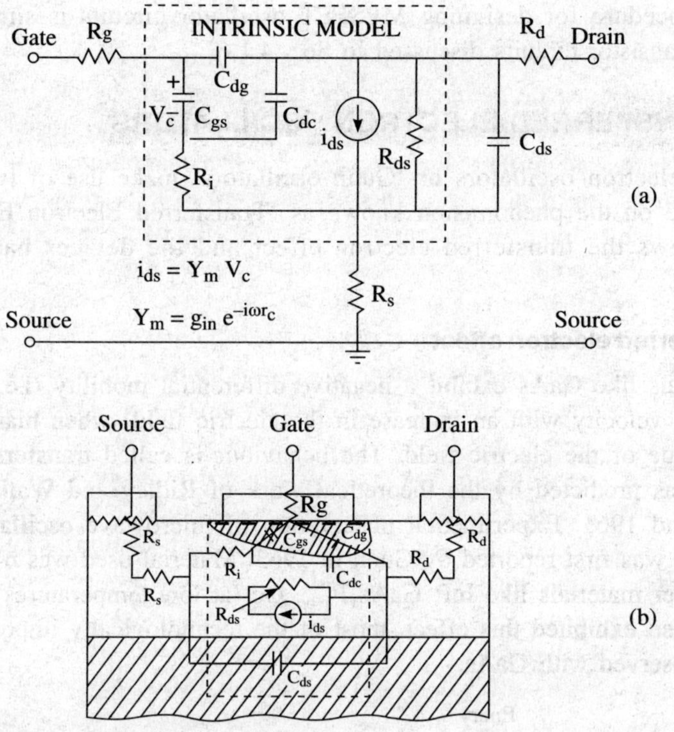

Fig. 4.11

The extrinsic (parasitic) elements are : R_s the source resistance, R_d the drain resistance, R_g the gate-metal resistance, and C_{ds} the substrate capacitance. Typical element values for a GaAs MESFET with 1 μm gate length and 500 μm gate width are listed in Table 4.1.

TABLE 4.1

Typical Values of Equivalent Circuit Parameters

Intrinsic Elements	Extrinsic Elements
g_m = 53 mmho	C_{ds} = 0.12 pF
τ_0 = 5.0 ps	R_g = 2.9 Ω
C_{gs} = 0.62 pF	R_d = 3 Ω
C_{dg} = 0.014 pF	R_s = 2.0 Ω
C_{dc} = 0.02 pF	*Bias Values*
R_i = 2.6 Ω	V_{DS} = 5 V
R_{ds} = 400 Ω	V_{GS} = 0
	I_{DS} = 70 mA

The procedure for designing MESFET oscillator circuits is similar to that for bipolar transistor circuits discussed in Sec. 4.1.6.

4.3 TRANSFERRED ELECTRON OSCILLATORS

Transferred electron oscillators or 'Gunn oscillators' make use of two terminal devices based on the phenomenon known as 'Transferred Electron Effect'. This section reviews the transferred electron effect and the devices based on this principle.

4.3.1 Transferred electron effect

Some materials like GaAs exhibit a negative differential mobility (i.e. a decrease in the carrier velocity with an increase in the electric field) when biased above a threshold value of the electric field. The behaviour is called transferred electron effect and was predicted by the theoretical work of Ridley and Watkins, and of Hilsum around 1961. Experimental observation of microwave oscillations based on this effect was first reported by Gunn in 1963. Material used was n-type GaAs. Although other materials like InP, $GaAs_xP_{1-x}$, Ge (at low temperatures), CdTe and ZnSe have also exhibited this effect, most of the technologically important results have been observed with GaAs.

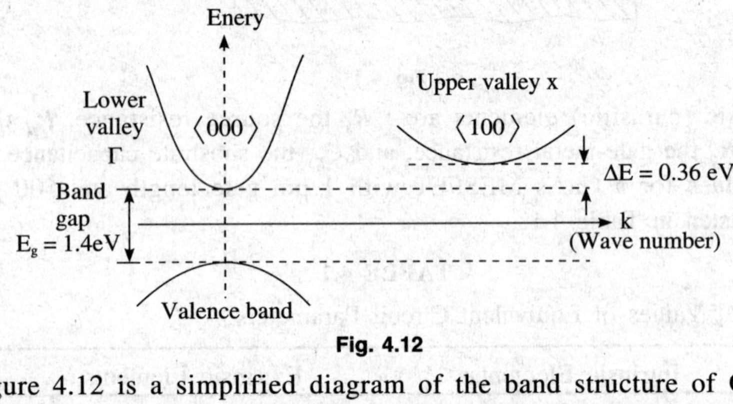

Fig. 4.12

Figure 4.12 is a simplified diagram of the band structure of GaAs. The conduction band is represented by two valleys, with the characteristics given in Table 4.2.

TABLE 4.2

Characteristics of the Conduction Band in GaAs

	Lower Valley (Γ)	Upper Valley *(X)*
Wave vector direction	< 000 > minimum	Collective effect of < 100 > minima
Effective mass	0.072 m_e	0.35 m_e
Mobility	8000 $cm^2\ V^{-1}s^{-1}$	150 $cm^2V^{-1}s^{-1}$

At zero voltage, the average energy of electrons is the thermal energy kT (0.025 eV at room temperature). Since this energy is less than the energy gap ΔE between two valleys in the conduction band, most of the electrons are in the lower valley Γ. As the voltage is increased, the electrons are accelerated by the electric field. The average carrier velocity v_0, and consequently, the kinetic energy increases. This increase in energy causes a redistribution of electrons between Γ and X valleys, more electrons being transferred to the upper valley X as the energy increases. When the energy gained by the electrons from the applied electric field is comparable to ΔE the rate of transfer into X valley (with lower mobility) increases. This transfer of electrons from the Γ valley into the X valley gives rise to a velocity versus field characteristics shown in Fig. 4.13.

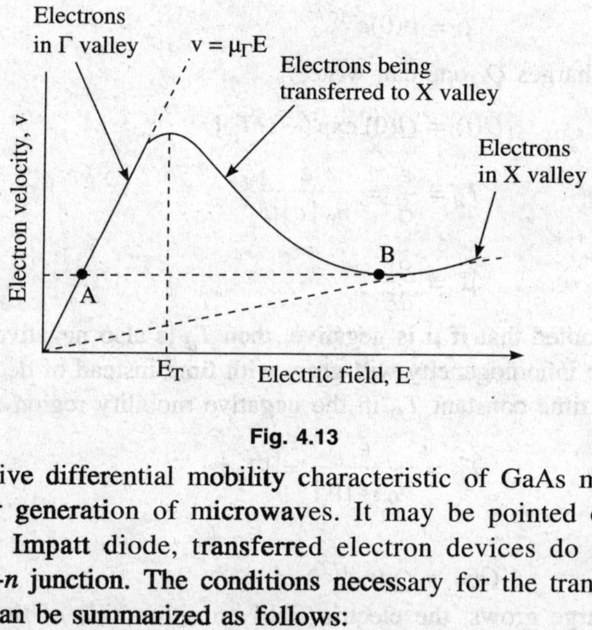

Fig. 4.13

This negative differential mobility characteristic of GaAs may be used for amplification or generation of microwaves. It may be pointed out that, unlike tunnel diode or Impatt diode, transferred electron devices do not require the presence of a *p-n* junction. The conditions necessary for the transferred electron effect to occur can be summarized as follows:

(i) At least two conduction valleys must exist with energies E_Γ and E_X such that $(E_X - E_\Gamma)$ is greater than kT_0.

(ii) E_Γ electrons must have a low effective mass and a high mobility, while E_X electrons must have a high effective mass and a low mobility.

(iii) $(E_X - E_\Gamma)$ should be much less than E_g (the band gap), so that the electric field required for electron transfer is much smaller than the breakdown field.

Electron dynamics in negative-differential mobility medium : The shape of the *v-E* characteristic of GaAs suggests that devices using a slice of this material with ohmic contacts on the top and bottom may be used directly as negative resistance amplifiers. Difficulty in utilizing these devices in this manner arises from the internal instabilities exhibited by a medium with a negative differential mobility. These instabilities are discussed below.

In a medium with positive differential mobility, any space charge inhomogeneity Q decays exponentially. This may be illustrated by interpreting Maxwell's equations discussed in Sec. 2.2.1. We have (with usual notations):

$$\nabla \times \mathbf{H} = \frac{\partial \mathbf{D}}{\partial t} + \mathbf{J}$$

Taking divergence and noting that the divergence of the curl of a vector is zero, we may write

$$0 = \frac{\partial}{\partial t}(\nabla \cdot \mathbf{D}) + \nabla \cdot \mathbf{J} = \frac{\partial \rho}{\partial t} + \nabla \cdot \sigma \mathbf{E} = \frac{\partial \rho}{\partial t} + \frac{\sigma}{\varepsilon}\rho \quad \text{...(4.15)}$$

A solution of (4.15) may be written as

$$\rho = \rho(0) e^{-(\sigma/\epsilon)t} \quad \text{...(4.16)}$$

or, in terms of charges Q, one can write,

$$Q(t) = Q(0) \exp[-t/T_d] \quad \text{...(4.17)}$$

where

$$T_d = \frac{\varepsilon}{\sigma} = \frac{\varepsilon}{n_0 |e| \bar{\mu}} \quad \text{...(4.18)}$$

and

$$\bar{\mu} = \frac{\partial v}{\partial E}$$

It may be noted that if μ is negative, then T_d is also negative. Consequently, any space charge inhomogeneity will grow with time instead of decaying. We shall define a growth time constant T_D in the negative mobility region as,

$$T_D = \frac{\varepsilon}{n_0 |e| |\bar{\mu}|} = |T_d| \quad \text{...(4.19)}$$

and therefore,

$$Q(t) = Q(0) e^{t/T_D} \quad \text{...(4.20)}$$

As the charge grows, the electric field on either side of the inhomogeneity builds up. This continues as long as the value of the electric field lies in the negative differential mobility region. As shown in Fig. 4.14, growth of the charge fluctuations causes a localized enhancement of the electric field. Increase in the electric field decreases the velocity further. This results in accumulation and depletion regions as shown in Fig. 4.14c. A high field domain is thus formed. A steady state will be reached when the field inside the domain corresponds to point B and the field outside the domain has reached point A in Fig. 4.13. Thus all the electrons will be travelling with a uniform velocity (of the order of 10^7 cm/s in GaAs). In these qualitative arguments, effect of the diffusion has been ignored.

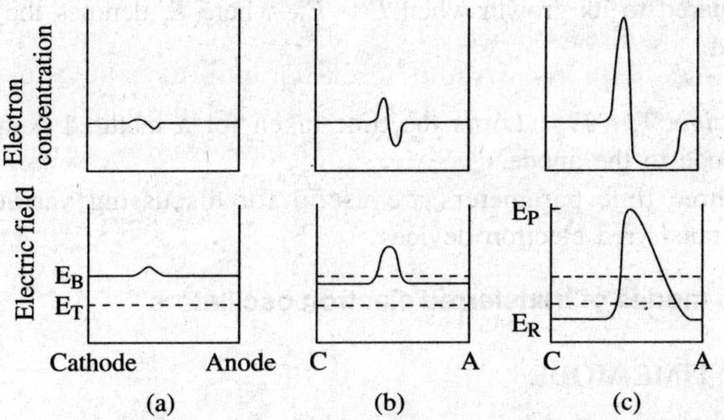

Fig. 4.14

Formation of high field domains requires that the transit time of the electrons through the GaAs sample should be larger than the domain growth time T_D. This condition may be written as,

$$\frac{L}{v} > T_D \text{ or } n_0 L > \frac{\varepsilon v}{e|\bar{\mu}|} \quad ...(4.21)$$

On the basis of a more accurate numerical analysis, one can write this criterion as $n_0 L > 10^{12}$ cm^{-2}. Devices which satisfy this inequality are called *supercritical,* while devices with $n_0 L$ less than 10^{12} cm^{-2} are called *subcritical.*

Normally, high field domains get nucleated near the cathode contact. The cathode contact forms a large discontinuity in doping, and the space charge build-up starts there. This high field domain drifts gradually during its growth and continues moving till it is collected at the anode. As soon as the domain is collected by the anode, the electric field across the sample becomes momentarily uniform. Immediately another high field domain starts growing near the cathode end and the process repeats itself.

TIME PARAMETERS FOR TEDs

There are three important time parameters for TEDs.

(a) *Domain-growth time constant* T_D : This time constant is the dielectric relaxation time in the negative mobility region

$$T_D = \frac{\varepsilon}{n_0 |e| |\bar{\mu}|} \quad ...(4.19)$$

The value of T_D is field dependent.

(b) Dielectric relaxation time in the positive mobility regime T_d :

$$T_d = \frac{\varepsilon}{n_0 |e| \mu_\Gamma}; \mu_\Gamma \simeq 7000 - 8000 \text{ cm}^2 V^{-1} s^{-1} \quad ...(4.22)$$

$T_d << T_D$ which means that the decay of the charge is much faster (when $E < E_T$) compared to its growth when $E > E_T$, where E_t denotes the value of the threshold field.

(*c*) *Transit time* T_t : $T_t = L/v$ is the time taken for a matured domain to travel from the cathode to the anode.

These three time parameters are useful for discussing various modes of operation of transferred electron devices.

4.3.2 Various modes of transferred electron oscillators

1. TRANSIT TIME MODE

This mode exists when the device is operated in a low Q circuit at a bias voltage just above the threshold voltage. Stable domains are formed. The period of oscillation is equal to the time taken for a domain to drift from its point of nucleation (usually the cathode contact) to the anode. As soon as the domain is collected at the anode, the voltage swings over the threshold level and the cycle repeats. The current in external circuit has a waveshape roughly as in Fig. 4.15.

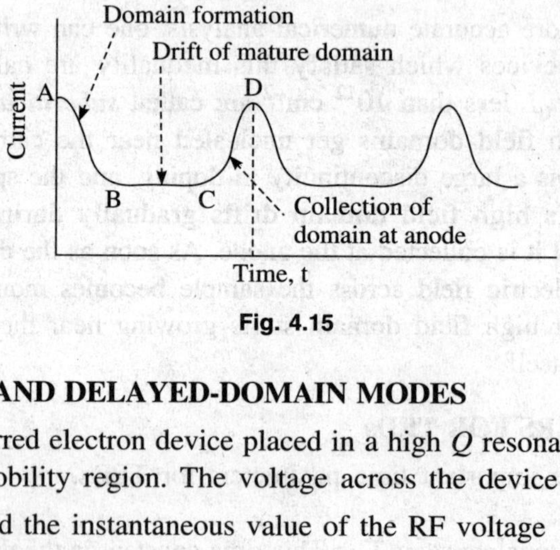

Fig. 4.15

2. QUENCHED AND DELAYED-DOMAIN MODES

Consider a transferred electron device placed in a high Q resonant cavity and biased in the negative mobility region. The voltage across the device *V(t),* is the sum of bias voltage V_b and the instantaneous value of the RF voltage $V_1 \cos(\omega_0 t + \phi)$. The total voltage can be written as,

$$V(t) = V_b + V_1 \cos(\omega_0 t + \phi) \qquad ...(4.23)$$

where ω_0 is the resonant frequency of the cavity, corresponding to a period $T_0 = 2\pi/\omega_0$. If V_1 is comparable to V_b, the total voltage $V(t)$ can swing below the threshold voltage during parts of the cycle and can either cause the domain to collapse or prevent the nucleation of a new domain. The following two modes are possible.

(a) *Quenched domain mode* $T_D < T_0 < T_t$: Since T_0 is greater than T_D, a domain that nucleates at the cathode when the voltage swings over the threshold value has sufficient time to mature. However, since T_0 is less than T_t, the voltage will swing below V_T before the domain reaches the anode. V_T is the threshold voltage corresponding to the threshold field E_T. When the voltage swings below the domain sustaining voltage V_s (which is less than V_T), the domain will be quenched even before it reaches the anode. As noted above, T_d is less than T_D and consequently T_d is much less than T_0, i.e., the quenching time constant is small compared to the period of oscillation. In the quenched mode of operation the life time of the domain is a function of the bias, the *RF* swing (and therefore the circuit *Q*), and the relationship between T_0 and T_t. Various voltage levels and time relationship are shown in Fig. 4.16.

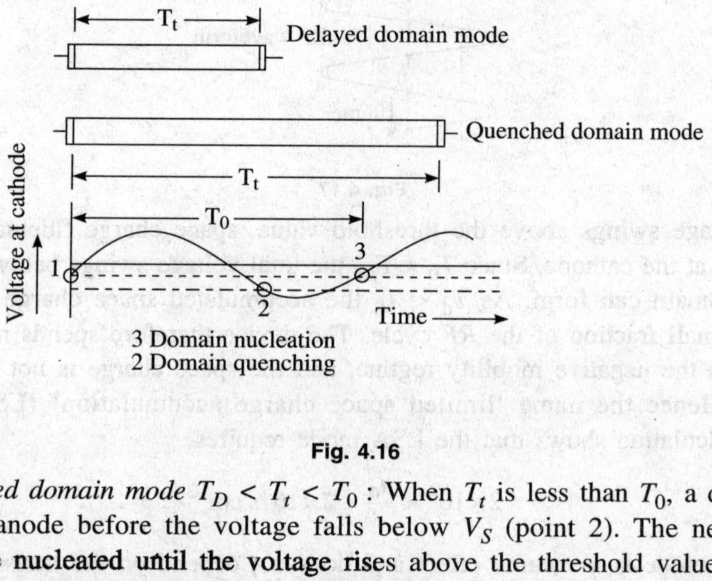

Fig. 4.16

(*b*) *Delayed domain mode* $T_D < T_t < T_0$: When T_t is less than T_0, a domain can reach the anode before the voltage falls below V_S (point 2). The next domain will not be nucleated until the voltage rises above the threshold value (point 3). The resonator therefore 'delays' the formation of the domains and this mode is called the inhibited or the delayed domain mode.

Thus, by choosing a resonant circuit for quenched or delayed domain mode, the *TEO* can be made to operate both below and above the intrinsic transit time frequency.

In all the three modes discussed above the frequency of oscillation is related to the transit time, and hence the physical dimension (thickness of the device). Since the devices operate at fixed value of electric field, the voltage across the device decreases linearly as the design frequency is increased. Also, as the device capacitance increases with decrease in thickness, the impedance of the device also decreases with frequency. Because of these considerations product of output power and impedance falls off as $1/f^2$. This is a characteristic feature of all transit time devices.

3. LIMITED SPACE CHARGE ACCUMULATION MODE ($T_d << T_0 < T_D$)

In the LSA mode, the device is placed in a resonator tuned to the frequency $f_0 (= 1/T_0)$, where f_0 is several times well above the transit frequency. The device is biased well above the threshold voltage. Relative amplitudes of the bias and the RF voltage are shown in Fig. 4.17.

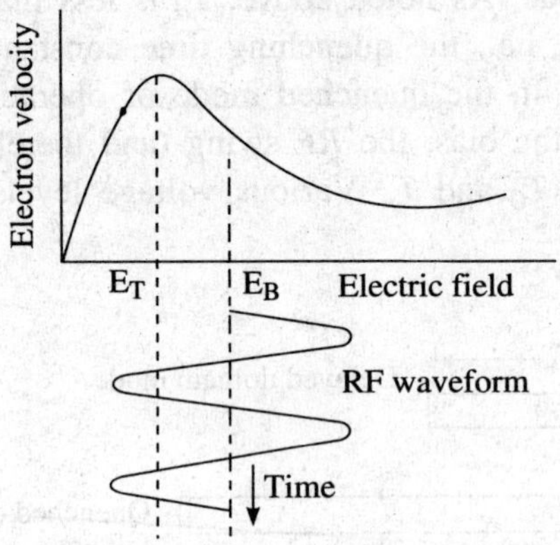

Fig. 4.17

As the voltage swings above the threshold value, space charge fluctuation starts building up at the cathode. Since $T_0 < T_D$, the total voltage swings below threshold before a domain can form. As $T_d < T_0$ the accumulated space charge is drained in a very small fraction of the *RF* cycle. The device therefore spends most of the RF cycle in the negative mobility regime, and the space charge is not allowed to build up. Hence the name 'limited space charge accumulation' (LSA) mode. Detailed calculation shows that the LSA mode requires,

$$2\times10^4 < \frac{n_0}{f} < 2\times10^5 \text{ s cm}^{-3} \qquad ...(4.24)$$

Since LSA mode is a volume effect and does not depend on the transit time, the power-impedance product does not fall off as $1/f^2$. Therefore, this mode is of potential advantage at high frequencies and for very high output powers.

4.3.3 Gunn oscillator circuits

Gunn diode oscillator circuits comprise of a resonant cavity, an arrangement for coupling diode to the cavity, a circuit for biasing the diode and a mechanism to couple the RF power from the cavity to the external circuit or load. A variety of circuits has been suggested for performing these functions. Two commonly used circuits use a coaxial cavity and a rectangular waveguide cavity, respectively.

CIRCUIT USING COAXIAL CAVITY

A typical configuration for this type of circuit is shown in Fig. 4.18. The diode is mounted at one end of the cavity and is in continuation with the central conductor of the coaxial line. The output can be taken out through a capacitively or inductively coupled probe. The frequency of oscillation is determined mainly by the length of the cavity. The location of the coupling loop (or probe) within the resonator determines the load impedance presented to the diode. Heat sink is included to conduct away the heat generated by the power dissipated in the device.

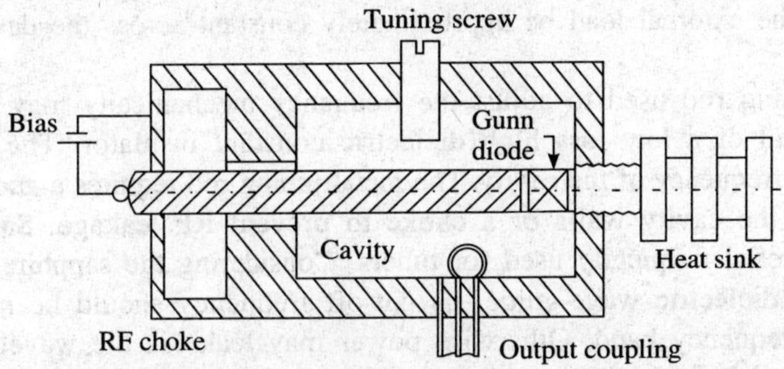

Fig. 4.18

An advantage of this circuit is that it can be fabricated relatively easily and the coupling may be adjusted on production units to tailor them to individual diodes. There are some disadvantages as well. The inherent Q of the coaxial resonator structure is low. Again, as a coaxial cavity is also resonant at harmonics of the fundamental frequency, the diode may oscillate at harmonics of the desired frequency.

CIRCUIT USING WAVEGUIDE CAVITY

The most widely used Gunn oscillator circuit is the post coupled rectangular waveguide cavity. This configuration is shown in Fig. 4.19 and consists of a simple waveguide section, separated from the output waveguide by an iris. The diode is mounted in a post across the narrow dimension in the centre of the guide.

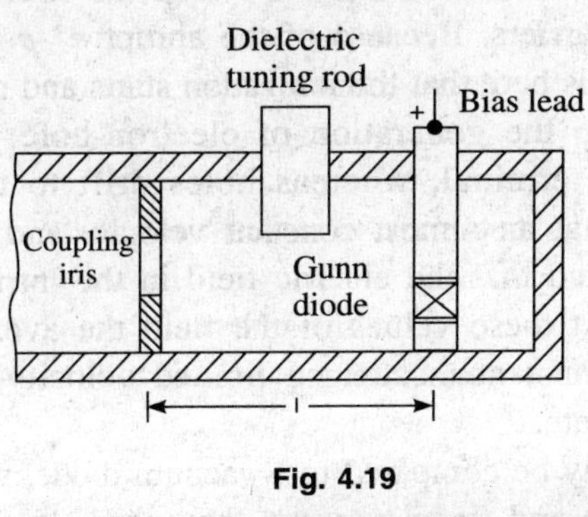

Fig. 4.19

This cavity operates in a TE_{101} mode. If the diode post were not present, the resonant frequency would be approximately such that the length of the cavity equals one-half-wavelength ($\lambda_g/2$). However, in the circuit shown in Fig. 4.19, the diode post acts as a large inductive susceptance and the iris is also inductive. As a result, the resonant frequency is lower than that for which the length 'l' is one-half guide wavelength.

For a broadband operation several other factors must be considered. The first requirement is that the cavity be tunable mechanically and also that the frequency change linearly with the depth of plunger. The second requirement is that the coupling to the external load be approximately constant across the desired tuning range.

The tuning rod used to adjust the frequency mechanically may be either a low loss metal or a low loss high dielectric constant insulator. The tuning rod decreases the frequency of the cavity. The metal tuning rod requires a good electrical contact with the cavity walls or a choke to prevent RF leakage. Sapphire is a dielectric material frequently used for tuners. Considering the sapphire tuning rod as a circular dielectric wave-guide, its cut-off frequency should be much above the desired frequency band. Otherwise power may leak via the waveguide mode in the tuning dielectric rod.

The coupling to the external load, determined by the iris size, usually changes with the frequency of operation. An iris designed for critical coupling (maximum power) at a frequency f_0, will undercouple the cavity at lower frequencies and overcouple at higher frequencies.

4.4 AVALANCHE DIODE OSCILLATORS

4.4.1 Avalanche diodes (or IMPATT diodes)

The history of avalanche diodes goes back to 1958 when W.T. Read proposed a high frequency negative resistance diode. He suggested a $n^+—p— i —p^+$ structure. This configuration is now known as Read's structure and is shown in Fig. 4.20.

The superscript + indicates a heavy doping and i stands for intrinsic (meaning an equal number of available holes and electrons). The structure is reverse biased as shown and the applied bias is sufficient to sweep the entire region between n^+ and p^+ terminals free of carriers. Because of the abrupt n^+-p junction, the electric field is highest at $x = 0$. It is here that the ionization starts and avalanche breakdown takes place. This leads to the generation of electron-hole pairs. Electrons are collected at the positive terminal, whereas holes drift to the right across the remaining depletion layer at an almost constant velocity and are collected at the p^+ terminal. It may be noted that the electric field in the intrinsic region is about 5000 volts/cm or more. At these values of the field the average carrier velocity becomes approximately constant (scattering limited velocity) and the term drift velocity becomes appropriate.

The situation here may be compared to a vacuum diode, wherein the electrons are emitted by the cathode and travel towards the anode. In a $p^+—n—i—n^+$ Read structure, electrons are generated near the p^+ electrode and travel towards the n^+ region.

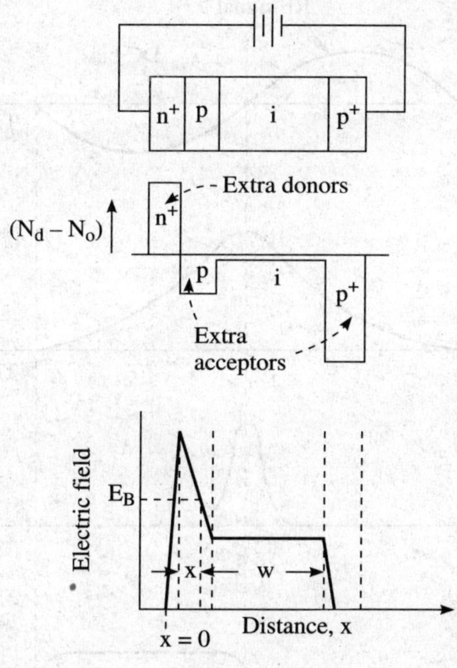

Fig. 4.20

To understand the working of a Read diode let us assume that the bias is adjusted such that E_0 is just below the critical field for avalanche breakdown. If an RF voltage is superimposed over this bias, the total field will go above the requirement for avalanche breakdown during the positive half cycle. Throughout this interval an exponential increase of the avalanche charge continues. During the negative half cycle the total field is below this critical value and the avalanche process decays. Since the growth rate is exponential, we have peak charge (in the avalanche region near $x = 0$) at the end of the positive half of the AC cycle. This build up of the avalanche charge is shown in Fig. 4.21. We note that there is a phase delay of approximately $\pi/2$ radians before the bunch of positive charge carriers drifts across the sample from $x = 0$ to p^+ region shown in Fig. 4.20. A charge moving at constant velocity within an electric field would generate a constant current in the external circuit. The finite transit time of the carriers results in a current pulse of a duration equal to the transit time. If the width of the *i*-layer is such that the transit time through is $T/2$, the current in the external circuit has a wave shape as shown in Fig. 4.21(d). As seen in these figures there is a phase difference of π between the voltage and the current. This implies that the resistance of the device is negative.

Based on these arguments, Read pointed out that one could design a high frequency negative resistance diode. The width of the depletion layer W would be such that transit time τ is half the time period at the frequency of operation.

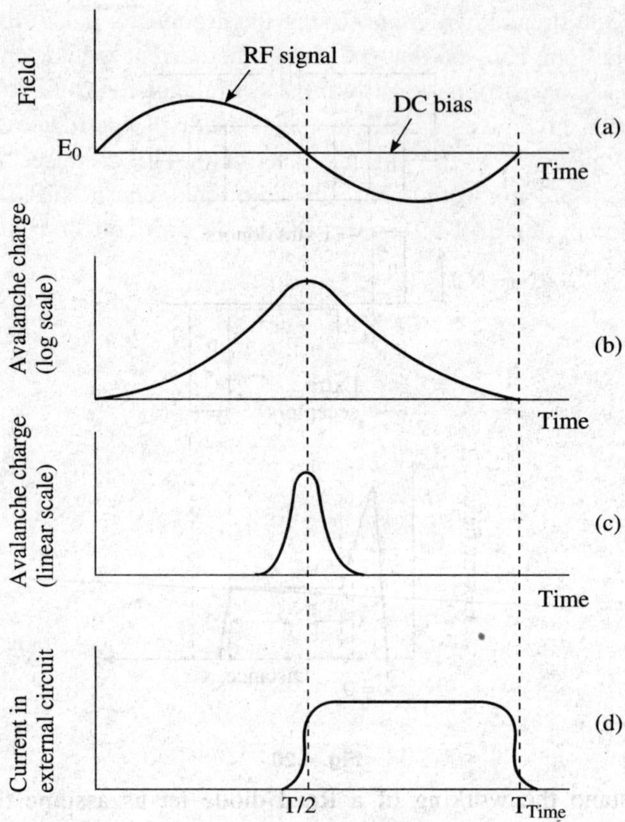

Fig. 4.21

Scattering limited velocity in silicon is about 10^7 cm/sec. This means that for operation at 5 GHz the width of the *i*-layer should be 10 microns.

In 1958, when this proposal was put up by Read, the silicon technology had not advanced enough and the configuration ($n^+ - p - i - p^+$) could not be fabricated to verify the proposal. It was seven years later (1965), that Johnston, DeLoach and Cohen realized a negative resistance device in silicon which could give 60 mW of pulsed power at 12 GHz. But surprisingly this device did not use Read's $n^+ — p — i — p^+$ (or $p^+ — n — i — n^+$) configuration but consisted of a simple reverse biased *p-n* junction. By this time silicon planar technology had developed very well, and soon the Read diode was also fabricated and the principle of operation was verified by Lee (at 200 MHz) and also by DeLoach and Johnston (at 5 GHz).

However, the results obtained by using a simple *p-n* junction indicated that the complicated configuration proposed by Read was not necessary. At first sight, it looks surprising that how any *p-n* (or *p-i-n*) diode could have similar characteristics as Read diode. However, when a *p-n* (or *p-i-n*) diode is reverse biased, avalanche breakdown takes place in the depletion layer, wherever the electric field exceeds the breakdown value. Using the arguments put forth while discussing the Read diode, one can see that the avalanche current would lag the applied field by π/2 radians. The carriers constituting the avalanche current from different parts of the depletion layer would drift to the respective electrodes (holes to negative electrode and electrons to the positive electrode). The distances to be travelled by various carriers are not equal, but any additional phase shift (between 0 and π radians) caused by the drift of carriers makes the carriers contribute to the negative resistance.

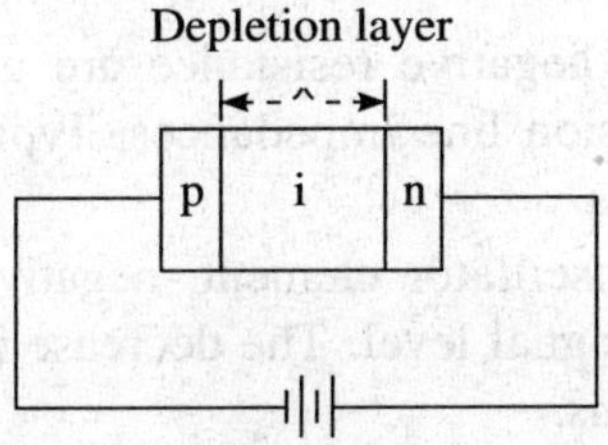

Fig. 4.22

A complete mathematical theory was soon put up by Misawa (1966) to consider avalanche transit time phenomenon in a *p—i—n* structure (Fig. 4.22). Essentially, this involves the solution of Poisson's equation taking into account the generation of carriers caused by the impact ionization. From this equation, a relation between the field and the current is obtained and the terminal impedance written as

$$Y = \frac{AJ}{V} = AJ\left\{\int_0^L E\,dx\right\}^{-1} \quad ...(4.25)$$

where L is the length of the device.

The main results of this analysis are: (i) a resonance frequency is observed below which the diode susceptance is inductive and above which it is capacitive; and (ii) the real part of the diode impedance becomes negative at frequencies well below this resonance frequency and remains negative over octaves of the frequency range.

4.4.2 Equivalent circuit of IMPATT diodes

An equivalent circuit for the IMPATT diode chip is shown in Fig. 4.23. Here the active part of the diode (the chip, excluding the package) is shown as a negative resistance R_D in series with a reactance X_D. R_D includes the unavoidable parasitic

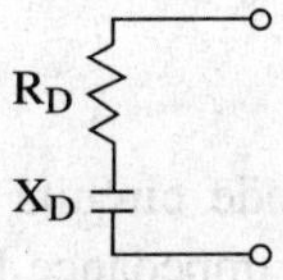

Fig. 4.23

series resistance and the resistance of the undepleted drift region (if any). Some important properties of this equivalent circuit are listed below:

(i) The magnitude of the net negative resistance R_D is usually much smaller than that of the reactance X_D. Consequently, the magnitude of the chip impedance is approximately X_D.

(ii) For most cases of interest, X_D can be approximated, with sufficient accuracy, by the reactance of the junction (chip) capacitance at the breakdown voltage. This capacitance ranges from 0.2 to 0.6 pF for typical diodes.

(iii) Values of the net negative resistance are generally small compared to the usual transmission line impedances. Typical values range from – 0.7 ohms to –2.0 ohms.

(iv) As in any active oscillator element, negative resistance of an IMPATT diode varies with signal level. The decrease in R_D with signal level leads to stable oscillations.

PACKAGED DIODES

To a good approximation, the package can be described by two reactive elements, a series inductance and a shunt capacitance. This holds good for packages used for Gunn diodes also. Equivalent circuit for a packaged IMPATT diode is shown in Fig. 4.24. The exact values of L_p and C_p vary somewhat from one package style to another but typical values are 0.6 nh and 0.3 pF, respectively. The value of L_p is also affected by the microwave circuit surrounding the diode. Typical values for a commercially available diode are also shown in Fig. 4.24. The value of R_D corresponds to an output power of about 700 mW as an oscillator. We note that the package inductance very nearly resonates the chip capacitance, so that the reactance presented by the chip at the package terminals is relatively small. The overall impedance of the packaged diode Z_{DP} for this case is – 2.7 + *j*8.0 ohms at 10 GHz.

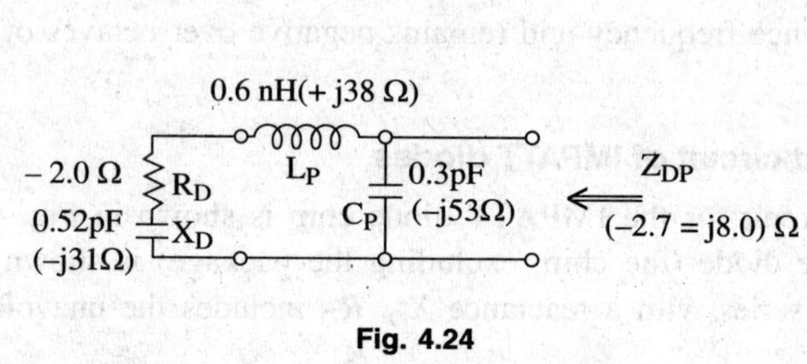

Fig. 4.24

4.4.3 Circuit considerations

Design philosophy for IMPATT diode circuits is similar to that for Gunn diode circuits. The only difference is in the impedance level of the two devices; impedance of Gunn devices is typically tens of ohms while for IMPATT diodes it is only in ohms.

Various types of circuits have been reported for IMPATT diode oscillators and amplifiers. These can be classified in three groups; broadly tunable circuits, low Q circuits and high Q circuits.

BROADLY TUNABLE CIRCUITS

These ars very flexible circuits and are useful as laboratory tools for the study of IMPATT diode properties. These include a multiple slug tuned cavity, variable inductance tunable coaxial circuit and a tunable waveguide oscillator circuit.

A multiple slug tuned cavity is a coaxial cavity shown in Fig. 4.25. *A* and *B* arc two low characteristic impedance (10 to 20 ohms) regions formed by inserting metallic sleeves, called slugs, over the inner conductor. Location of these slugs can be changed by sliding them over the central conductor. Length of the slugs is normally kept equal to a quarter wave-length at centre frequency. By adjusting the position of the slugs it is possible to match diodes of various impedances. When the distance between the first slug B and the shorted end of the coaxial line *S* is approximately half wavelength, this region behaves as a resonator and the two slugs control the coupling between the cavity and the output line. The impedance transformation produced by this coupling arrangement can be varied by changing the impedance and the location of the slugs.

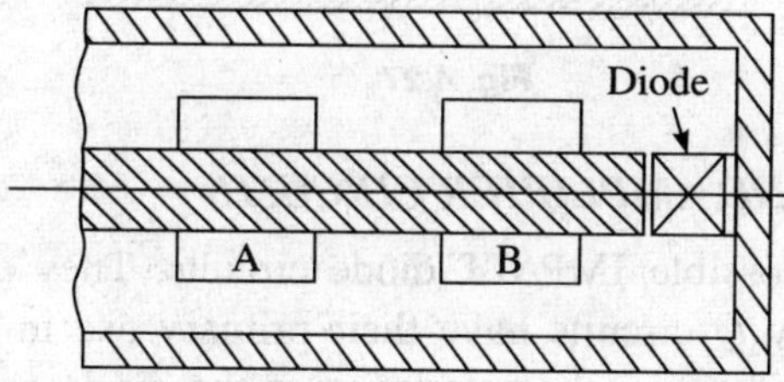

Fig. 4.25

A variable inductance tunable coaxial circuit is shown in Fig. 4.26. In this circuit the package inductance is used to resonate the chip capacitance. A lumped equivalent circuit is also shown in this figure. The frequency of oscillation of this circuit is changed by varying L_p. Package inductance L_p is not unique to a given package, and can be varied by changing the geometry of the circuit surrounding the diode. The diode package is end-mounted in a coaxial line in such a way that the package can be recessed into the ground plane in an adjustable manner. The dimension x shown in Fig. 4.26 is made variable. Recessing the diode decreases the inductance. A variation of say 0.4 nh to 0.6 nh can be obtained by a recession of the order of 0.050 inches. Two-section quarter wave transformer shown in Fig 4.26 reduces the 50 ohm line impedance to the value of R_D (usually in the vicinity of 2 ohms).

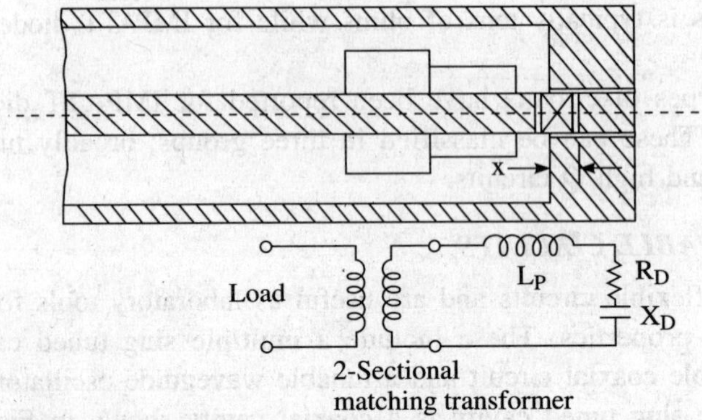

Fig. 4.26

Figure 4.27 shows a schematic illustration of a tunable waveguide oscillator. A standard sliding short can be used for tuning. For smaller tuning range requirements, the sliding short can be replaced by a fixed short and a tuning screw.

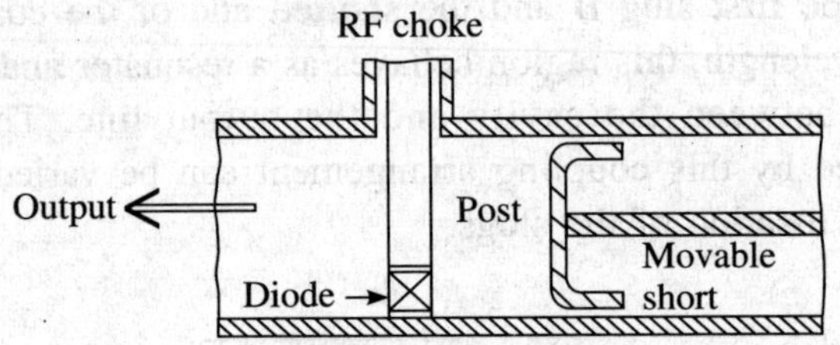

Fig. 4.27

LOW Q OSCILLATOR AND AMPLIFIER CIRCUITS

These are the simplest possible IMPATT diode circuits. They do not possess any frequency tunability. Low *Q* circuits have their primary use in applications where the frequency stability and noise characteristics of the diode can be controlled by injection phase locking techniques. Typical applications are: as a final output stage in an FM telecommunications transmitter, as an up-converter pump and as a CW Doppler radar transmitter.

Circuit configurations for these oscillators include: (i) coaxial circuits with suitable transformation from the diode to the transmission line impedance and (ii) waveguide circuit with the diode mounted on a post at a fixed distance from the shorted end.

HIGH Q LOW NOISE OSCILLATORS

The noise performance and frequency stability achievable with these circuits is compatible with local oscillator requirements in radar, telemetry and telecommunication receivers. Sufficient power is available from these oscillators to make them suitable for low power transmitters and for parametric amplifier pumps, as well as for police radars and intrusion alarms.

A coupled coaxial-waveguide oscillator is most commonly used for this class of oscillators. This circuit configuration is shown in Fig. 4.28. The design basically consists of a coaxial resonator which is strongly coupled to a high Q waveguide resonator. The waveguide resonator is in turn coupled by an iris to the load. Coupling between the coaxial circuit and the waveguide resonator is through the magnetic field. This is achieved by passing the centre conductor of the coaxial oscillator circuit directly through the bottom (broad) wall of the waveguide resonator at a position corresponding to the magnetic field maximum in the resonator. The centre conductor of the coaxial line is extended further through the top wall and passes through a matched load to the bias connection. The function of the matched coaxial load is to prevent spurious resonances in the coaxial portion of the oscillator.

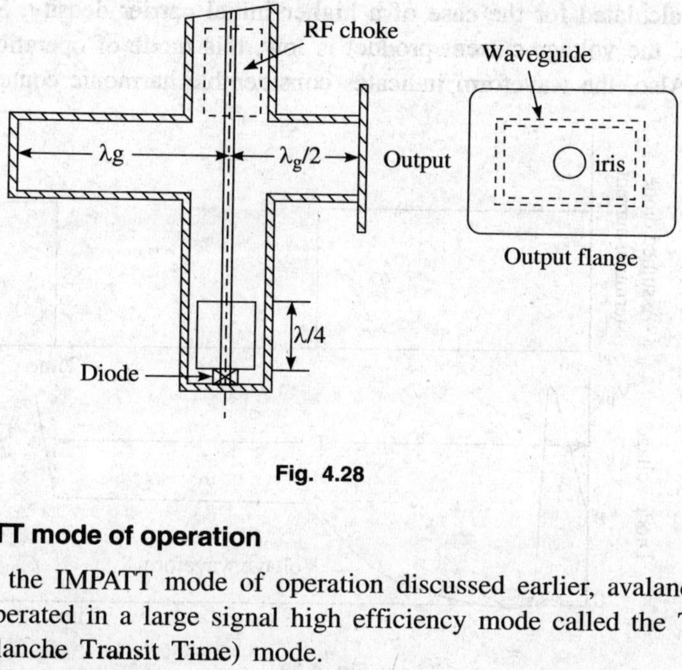

Fig. 4.28

4.4.4 TRAPATT mode of operation

In addition to the IMPATT mode of operation discussed earlier, avalanche diodes can also be operated in a large signal high efficiency mode called the TRAPATT (Trapped Avalanche Transit Time) mode.

Before discussing this mode of operation let us briefly recall the IMPATT behaviour. Two key phenomena taking place in the IMPATT mode are: (i) carrier generation by impact ionization producing a current pulse with a phase delay of 90°, and (ii) an additional phase shift introduced by the drift of carriers. The TRAPATT oscillations also depend on the delay in the current caused by the avalanche process. In this case the avalanche delay makes it possible to drive the diode voltage well above the normal breakdown voltage (for example twice as much) provided the driving voltage has a very fast rise time. This application of the excess voltage is followed by a very rapid multiplication of carriers. The carrier concentration greatly exceeds that required to sustain the terminal current of the diode. Space charge, created by these excess carriers, causes the field in the avalanche region to collapse, leaving the region filled with a plasma of electrons and holes. After the creation of the plasma, the diode is in a near zero voltage and high current state. If the terminal current is maintained, the plasma is removed from the diode and the diode voltage recovers. Removal of plasma is a space charge limited process and the recovery time involved in this process is typically several times the transit time corresponding to saturated velocity. Also, if the stray capacitance in the immediate vicinity of the diode is discharged by the plasma extraction, the diode current must drop to a low value and the diode voltage remain constant (approximately at its breakdown value).

A simple analysis of the TRAPATT operation may be carried out by assuming that the diode terminal current is a square wave and then calculating the voltage response. For this case, the sequence of events mentioned above results in the diode terminal voltage shown in Fig. 4.29. The dotted line indicates the waveform calculated for the case of a higher initial carrier density. Since at most of the times, the voltage-current product is low, this mode of operation has a high efficiency. Also, the waveform indicates considerable harmonic content.

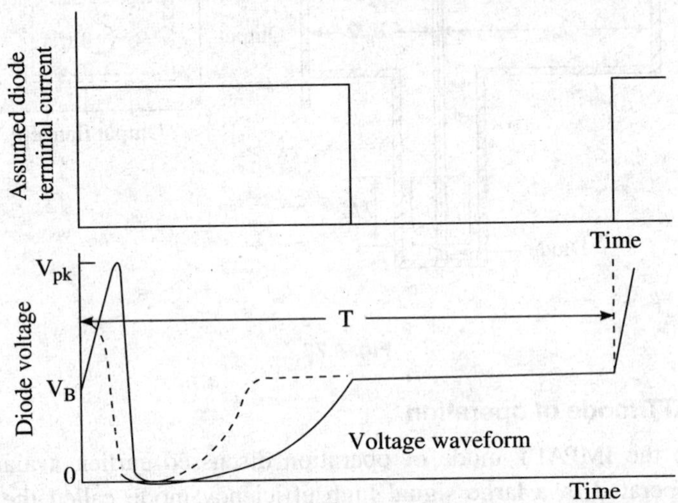

Fig. 4.29

A circuit configuration for the TRAPATT mode operation is shown in Fig. 4.30. The diode is mounted at one end of a transmission line. Diode package reactances are represented by a bond-wire inductance L and a package capacitance C. The other end of the line is terminated in a low pass filter placed approximately one half of a wavelength from the diode. Cut off frequency of the filter is just above the TRAPATT oscillation frequency. The TRAPATT operation interacts with the circuit in the following manner. As discussed earlier, the overdrive of the diode voltage is followed by a rapid collapse in the field. This results in a negative going voltage pulse which propagates from the diode down the line to the low-pass filter. The low-pass filter presents a short circuit to all the harmonics of the pulse and thus reflects a positive-going voltage back towards the diode. If this positive pulse is of sufficient amplitude, the diode will be quickly driven above its breakdown voltage and the sequence will repeat itself. Thus the circuit shown in Fig. 4.30 is a simple way of obtaining the desired triggering in the diode.

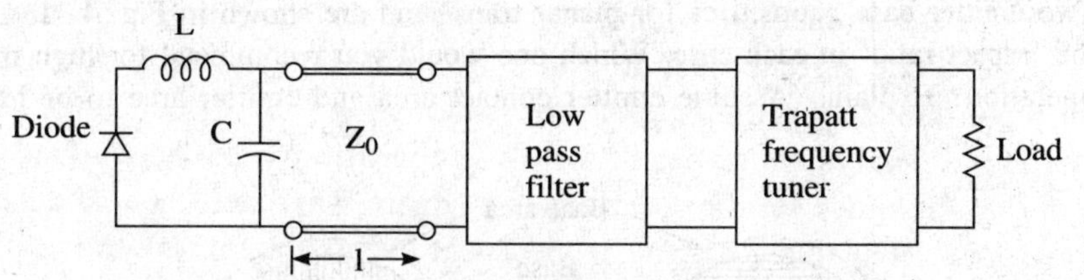

Fig. 4.30

TRAPATT devices have been operated at frequencies from 400 MHz to about 12 GHz. TRAPATTs are expected to successfully compete with other microwave devices for system applications which require high pulsed power outputs, high efficiency and moderate bandwidths. Some potential applications are phased-array radar systems, IFF transmitters (beacons), proximity fuse sources, radio altimeters and microwave landing systems.

SELECT BIBLIOGRAPHY

H. A. Watson (Ed), *Microwave Semiconductor Devices and their Circuit Applications,* McGraw-Hill Book Company, 1969. Chapter 15, Avalanche Transit-time Microwave Devices; Chapter 16, Bulk Gallium Arsenide Devices; Chapter 17, Microwave Transistors and Chapter 18, A 4-GHz Transistor Amplifier.

B.C. Deloach, Jr., 'Recent advances in solid state microwave generators', in *Advances in Microwaves,* Vol. 2, Academic Press, 1967.

H. Sobol and F. Sterzer, 'Microwave Power Sources', IEEE Spectrum 1972, pp. 20-33.

G.S. Hobson, *The Gunn Effect,* Oxford, Clarendon Press, 1974.

G. Gibbons, *'Avalanche Diode Microwave Oscillators',* Clarendon Press, Oxford, 1973.

E.T. Casterline and James A. Benjamin, 'Microwave Power Transistors: The present and the future', *Electronic Progress,* Vol. XV, No. 1, 1973.

Special issue on microwave semiconductors, 'Proceedings of the IEEE', Vol. 59. No. 8, Aug. 1971.

K.C. Gupta and A. Singh, *Microwave Integrated Circuits,* New Delhi, Wiley Eastern, 1974, Chapters 8 and 9, 'Active Devices for MICs'.

D.W. Atchley, 'Gunn diode circuit handbook', Microwave Associates/HB-9000/Feb., 1971.

Hewlett Packard, 'Microwave power generation and amplification using IMPATT diodes', Application Note No. 935.

C.A. Liechti, 'Microwave field-effect transistors—1976', *IEEE Trans. on Microwave Theory and Tech.,* Vol. **MTT—24**, June 1976, pp. 279-299.

PROBLEMS

4.1 In a bipolar transistor, the emitter charging time is given by equation 4.6 and constitutes 40 per cent of the total transit time. Calculate the percentage change in the cut-off frequency when the emitter current is increased from 100 mA to 200 mA. All other parameters may be assumed to remain constant.

4.2 Two emitter-base geometries for planar transistors are shown in Fig. 4.31. Calculate the 'aspect ratio' in each case. Which one would you recommend for high frequency operation? Explain. (Assume emitter contact area and emitter area to be identical.)

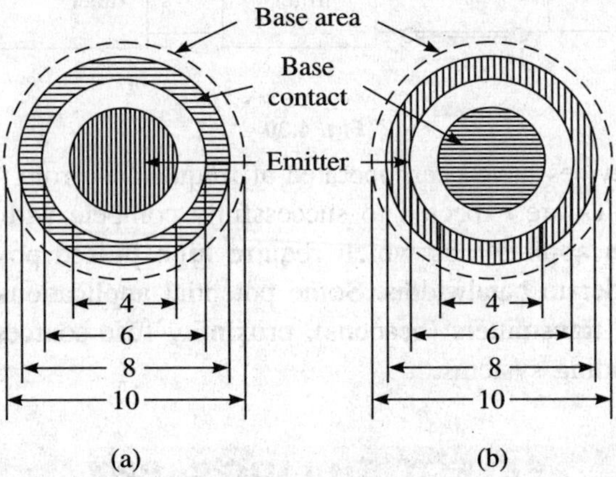

Fig. 4.31

4.3 A transferred electron diode operates in transit time mode at 10 GHz. If the velocity of propagation of domain is taken as 10^7 cm/sec, calculate the approximate thickness of the device. Can the device operate at 8.0 GHz and 12 GHz frequencies? Justify the modes of operation involved at these two frequencies.

4.4 Active impedance of a suitably biased TED is $(-10 + j7.5)$ ohms at 10 GHz. The diode is to be used in a coaxial line oscillator circuit similar to that shown in Fig. 4.18. Ignore the effect of tuning screw and the output coupling. Calculate the cavity length required when the characteristic impedance of the coaxial line used is 20 ohms. Smith chart may be used for calculations.

4.5 Impedance of a packaged avalanche diode at 10 GHz is $(-2.0 + j8.0)$ ohms. It is used in a coaxial circuit with only one quarter wave slug (*see* Fig. 4.25). Determine the spacing of the slug from the diode and the impedance of the slug so that the diode looks into an impedance which is negative of its own. Characteristic impedance of the coaxial line and the impedance of permitting load are 50 ohms.

CHAPTER 5

Microwave Network Representation

Microwave networks are often characterized in terms of their scattering matrix representation. The scattering matrix expresses a relationship between incoming and outgoing waves at various ports of a network. This is an alternative to impedance (or admittance) matrix representation commonly used at lower frequencies. Impedance and admittance matrices are based on Kirchhoff's laws for electrical networks. Kirchhoff's laws may be derived from Maxwell's equations.

5.1 KIRCHHOFF'S LAWS AND MAXWELL'S EQUATIONS

Kirchhoff's laws form the basis of network analysis in electrical engineering. In this section we shall discuss these two laws to point out their consistency with Maxwell's equations for electromagnetic fields.

Kirchhoff's first law states that *the algebraic sum of all currents flowing out of a junction must be zero.* In terms of notations shown in Fig. 5.1, we may write

$$\sum_{n=1}^{n} I_n = 0 \qquad \text{...(5.1)}$$

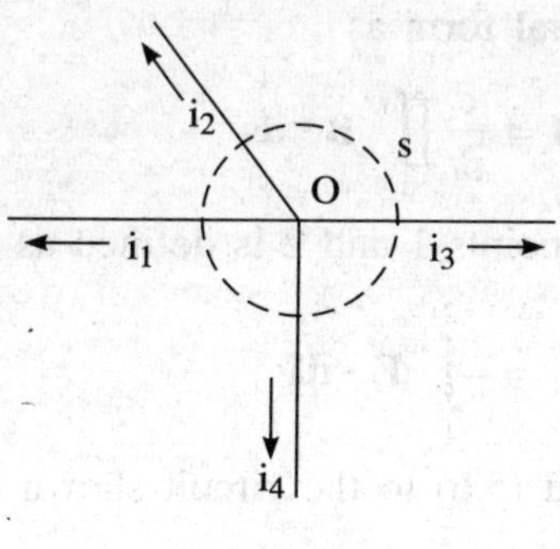

Fig. 5.1

The basic concept embodied in this law is that of the continuity of current. The continuity relation for electromagnetic fields may be derived by taking the divergence of curl H and may be written as

$$\nabla \cdot (\nabla \times \mathbf{H}) = 0 = \nabla \cdot \mathbf{J} + \frac{\partial}{\partial t}\Delta \cdot D = \nabla \cdot \mathbf{J} + \frac{\partial \rho}{\partial t} \quad ...(5.2)$$

This relation may be written in the integral form as

$$\iiint_v \nabla \cdot \mathbf{J}\, \mathbf{dv} + \frac{\partial}{\partial t} = \iiint_V \rho\, dv = 0 \quad ...(5.3a)$$

Using the divergence theorem for the first integral we have,

$$\oiint_S \mathbf{J} \cdot \mathbf{ds} = -\frac{\partial}{\partial t}\iiint_V \rho\, dv \quad ...(5.3b)$$

When (5.3b) is applied to the configuration shown in Fig. 5.1, the surface integral of current density may be written as a summation of currents and we have

$$\sum_{n=1}^{N} I_n = -dQ/dt \quad ...(5.4)$$

where Q is the charge stored within the volume enclosed by a surface S surrounding the node. Comparison of equations (5.1) and (5.4) shows that Kirchhoff's first law is consistent with Maxwell's equation provided there is no charge storage at junctions or nodes. This assumption is implicit in the formulation of Kirchhoff's first law and wherever charge storage takes place we add an additional branch or element to represent capacitive current dQ/dt.

Kirchhoff's second law states that *algebraic sum of voltages around a closed mesh equals zero*. This law may also be interpreted in terms of Maxwell's equations. As we shall see, this interpretation points out a limitation on the validity of the second law.

Consider the simple network shown in Fig. 5.2. We assume that all the energy storage and dissipation in the circuit occurs within the lumped elements Z_1, Z_2 and Z_3 which may contain inductors, capacitors and resistors. Losses and energy storage in the interconnecting leads would be neglected. The equation for curl E may be written in the integral form as

$$-\oint \mathbf{E} \cdot \mathbf{Dl} = \frac{\partial}{\partial t}\iint_S \mathbf{B} \cdot \mathbf{ds} \quad ...(5.5)$$

Also, voltage between two points 1 and 2 is defined as

$$V_{21} = -\int_1^2 \mathbf{E} \cdot \mathbf{dl} \quad ...(5.6)$$

Applying relations (5.5) and (5.6) to the circuit shown in Fig. 5.2, we may write

$$\sum_{n=1}^{N} V_n = \frac{\partial}{\partial t}\iint_S \mathbf{B} \cdot \mathbf{ds} \quad ...(5.7)$$

where the left hand side is the sum of voltage drops around a chosen path (the main rectangular loop in the figure). Since it has been assumed at the outset that all energy storage occurs in the lumped elements the magnetic field within the loop is zero. Therefore,

$$\sum_{n=1}^{N} V_n = -V_0 + V_1 + V_2 + V_3 = 0 \quad ...(5.8)$$

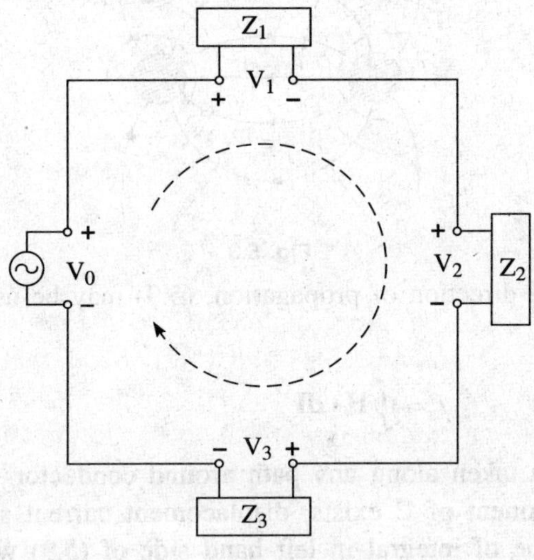

Fig. 5.2

Thus we see that Kirchhoff's second law is also consistent with Maxwell's equations provided there is no time-varying magnetic field in the neighbourhood of the circuit.

This discussion of Kirchhoff's laws helps us in understanding that basically Kirchhoff's laws and Maxwell's relations relate to same physical phenomena.

5.2 VOLTAGES AND CURRENTS

In order to make use of the low frequency circuit concepts at microwave frequencies, it is useful to describe the operation of microwave circuits in terms of **voltages and** currents. Voltage V and current I are related to electromagnetic fields **E and H** by the following relationships (which may be derived from **Maxwell's equations**),

$$\oint \mathbf{H} \cdot \mathbf{dl} = I + \iint_S \frac{\partial \mathbf{D}}{\partial t} \cdot \mathbf{ds} \quad ...(5.9)$$

and

$$\oint \mathbf{E} \cdot \mathbf{dl} = -\iint_S \frac{\partial \mathbf{B}}{\partial t} \cdot \mathbf{ds}$$

or

$$V_{BA} = -\int_A^B \mathbf{E} \cdot \mathbf{dl} = \iint_S \frac{\partial \mathbf{B}}{\partial t} \cdot \mathbf{ds} \quad ...(5.10)$$

Applying these relations to electromagnetic field of various modes in microwave transmission lines, one finds that the voltage and the current can be defined uniquely only for TEM mode on a two conductor line. Field configuration for a two-wire line is shown in Fig. 5.3. Since there no component of the electric

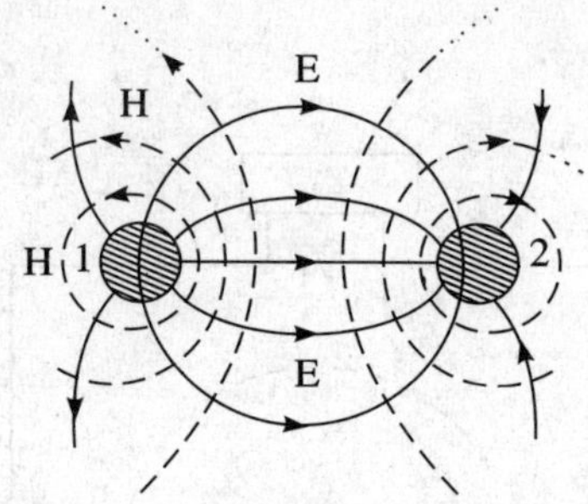

Fig. 5.3

flux density D in the direction of propagation, (5.9) may be used to define current as

$$I = \oint \mathbf{H} \cdot d\mathbf{l} \qquad \text{...(5.11)}$$

where the integral is taken along any path around conductor 1 or conductor 2. If a longitudinal component of E exists, displacement current should be taken into account and the value of integral in left hand side of (5.9) will depend upon the path of integration selected.

For a TEM wave the voltage V_{21} may be defined as

$$V_{21} = -\int_1^2 \mathbf{E} \cdot d\mathbf{l} \qquad \text{...(5.12)}$$

Since there is no longitudinal component of the magnetic field present, the integral in equation (5.12) is independent of the path of integration between conductors 1 and 2. However, if there is a longitudinal component of H, the integral in (5.12) does not have a unique value. Thus it becomes difficult to define uniquely voltages and currents for non-TEM modes.

Let us consider the formulation of voltage and current definitions for transverse electric (TE) and transverse magnetic (TM) modes in rectangular and circular waves guides. Waveguides have only one closed conducting boundary and it is difficult to decide which two points on this boundary should be selected for specifying the voltage. For TM waves, the line integral of (E · dl) between any two points becomes zero. Since there is no longitudinal component of magnetic field (perpendicular to the plane of paper), the line integral along the closed path shown in Fig. 5.4 should be equal to zero. Since the tangential component of the electric field along the wall is zero, its integral along the wall and, therefore, the integral along any two points A and B must also be zero. On the other hand, for

TE modes, there is a longitudinal component of magnetic field.

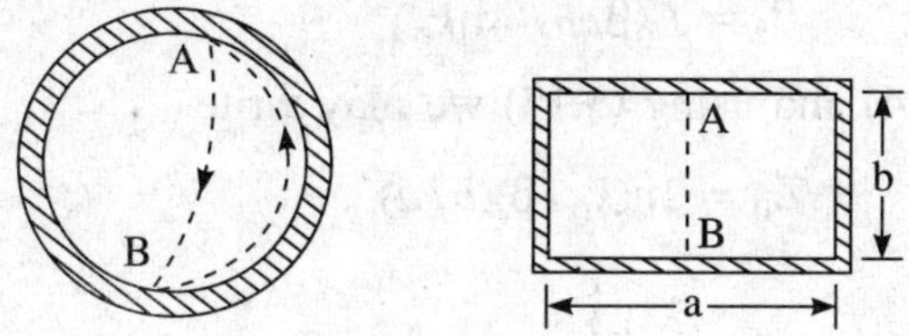

Fig. 5.4

Therefore, the integral along the path connecting A and B in Fig. 5.4 will not be zero but will depend upon the path selected. Consequently, voltage cannot be defined uniquely in this case also.

To conclude the discussion in this section, we note that (a) voltage and current can be defined uniquely for TEM modes only, and (b) for non-TEM modes the integrals giving voltage and/or current do not yield unique values.

5.3 WAVEGUIDE IMPEDANCE

Most commonly used waveguide mode is the TE_{10} mode in a rectangular waveguide. As discussed above, a unique voltage cannot be defined for TE modes. Nevertheless, it is useful to derive an expression for the waveguide impedance even in terms of a voltage which is defined in somewhat ad hoc manner. For a TE_{10} rectangular waveguide mode, a useful voltage definition is obtained by selecting the path AB midway and parallel to the two narrow walls as shown in Fig. 5.4. This voltage V_0 may be written as

$$V_0 = E_0 b \qquad ...(5.13)$$

where E_0 is the field in midplane and b is the height of the waveguide. Waveguide impedance may now be defined as

$$Z_0 = V_0^2 /(2P_z) \qquad ...(5.14)$$

where P_z is the power flowing in the waveguide. Power flow in z-direction may be calculated by using the Poynting theorem as follows :

$$P_z = \frac{1}{2}\iint -\text{Re}\left[E_y H_x^*\right] ds \qquad ...(5.15)$$

where the integral is evaluated over the cross-section of the waveguide. We may use equation 2.71 to write the fields for TE_{10} mode as

$$E_y = E_0 \sin\frac{\pi x}{a} e^{j\omega t} e^{-j\beta z} \qquad ...(5.16)$$

$$H_x = \frac{-E_0\beta}{\eta k_0} \sin\frac{\pi x}{a} e^{j\omega t} e^{-j\beta z} \qquad ...(5.17)$$

where $$\eta = \sqrt{\mu/\varepsilon} \text{ and } k_0 = \omega\sqrt{\mu\varepsilon} \qquad ...(5.18)$$

Substituting (5.16) and (5.17) in (5.15) we have

$$P_z = E_0^2 \beta ab / (4\eta k_0) \qquad ...(5.19)$$

Substituting it in (5.14) and using (5.13) we may write

$$Z_0 = 2\eta (k_0 / \beta)(b / a) \qquad ...(5.20)$$

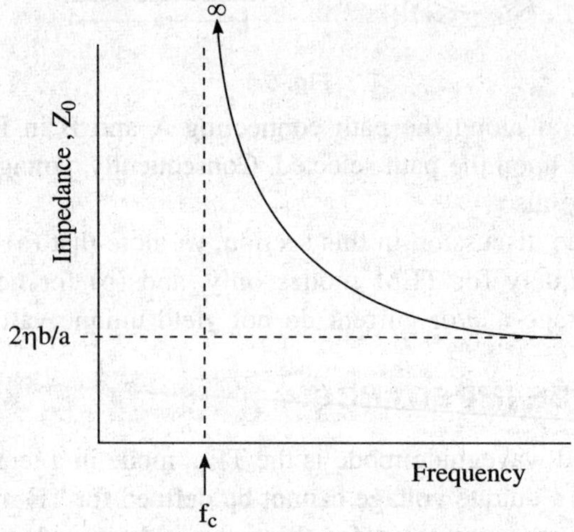

Fig. 5.5

Waveguide impedance as given by equation 5.20 is a function of frequency and waveguide dimensions. Variation with frequency is shown in Fig. 5.5. Impedance is infinite at the cut off frequency of the TE_{10} mode in the guide and tends to a value ($2\eta b/a$) at very high frequencies.

5.4 SCATTERING MATRIX REPRESENTATION

5.4.1 Definition

Representation of microwave networks by impedance or admittance matrices is not very convenient since voltages, currents and impedances cannot be measured in a direct manner at microwave frequencies. The quantities that may be measured easily and directly are reflection coefficient and transmission coefficient. These form the basis of the scattering matrix formulation. A scattering matrix represents the relationship between the parameters a_n's (proportional to the incident wave at nth port) and b_n's (proportional to the outgoing wave at nth port) defined in the following manner:

$$a_n = V_n^+ / \sqrt{Z_{on}} \qquad ...(5.21)$$

$$b_n = V_n^- / \sqrt{Z_{on}} \qquad ...(5.22)$$

where V_n^+ and V_n^- represent incident and outgoing waves along the line connected to the nth port and Z_{on} is the characteristic impedance of the line. Relationships between b_n's for a two-port network may be written as

$$b_1 = S_{11}a_1 + S_{12}a_2$$
$$b_2 = S_{21}a_1 + S_{22}a_2 \quad ...(5.23a)$$

or using matrix notation

$$[b] = [S][a] \quad ...(5.23b)$$

The matrix $[S]$ is known as the scattering matrix for the network. Coefficients S_{11} and S_{22} are reflection coefficients at ports 1 and 2, respectively, and the coefficient S_{12} and S_{21} represent transmission coefficients from port 2 to port 1 and from port 1 to port 2, respectively.

The average power flowing into the port n may be evaluated by using relations (5.21) and (5.22). Equations (2.11) and (2.15) may be used to write V_n and I_n as:

$$V_n = V_n^+ + V_n^- = \sqrt{Z_{on}}\,(a_n + b_n) \quad ...(5.24)$$

$$I_n = \frac{1}{Z_{on}}(V_n^+ - V_n^-) = \frac{1}{\sqrt{Z_{on}}}(a_n - b_n) \quad ...(5.25)$$

Power flow at the port n is given by

$$W_n = \frac{1}{2}\text{Re}(V_n I_n^*)$$
$$= \frac{1}{2}\text{Re}\{(a_n a_n^* - b_n b_n^*) + (b_n a_n^* - b_n a_n^*)\} \quad ...(5.26)$$
$$= \frac{1}{2}(a_n a_n^* - b_n b_n^*)$$

This shows that W_n is equal to the power carried into the port n by the incident wave (which is given by $\frac{1}{2}a_n a_n^*$) less the power reflected back (given by $\frac{1}{2}b_n b_n^*$).

5.4.2 Symmetry of S-matrix

The relations (5.21) and (5.22) permit us to establish a relationship between the scattering and the impedance matrices. We have

$$[V] = [Z] \cdot [I] \quad ...(5.27)$$

Using (5.24) and (5.25) and assuming that lines (or waveguides) connected to various ports have equal characteristic impedances denoted by Z_0, we get;

$$[V^+] + [V^-] = [Z]\frac{1}{Z_0}([V^+] - [V^-]) \quad ...(5.28)$$
$$= [Z']([V^+] - [V^-])$$

Where $$[Z'] = [Z]/Z_0 \quad ...(5.29)$$

Equation (5.28) may be written as

$$([Z'] + [U])[V^-] = ([Z'] - [U])[V^+] \quad ...(5.30)$$

where $[U]$ is the unity matrix. Since from equation (5.23)

$$[V^-] = [S][V^+]$$

and from equation (5.30)

$$[V^-] = ([Z']+[U])^{-1}([Z']-[U])[V^+] \quad ...(5.31)$$

we may write

$$[S] = ([Z']+[U])^{-1}([Z']-[U]) \quad ...(5.32)$$

An alternative form for the scattering matrix $[S]$ may be derived as follows. We have from equations (5.24) and (5.25)

$$[V^+] = \frac{1}{2}([V]+Z_0[I]) = \frac{1}{2}([Z]+Z_0[U])[I] \quad ...(5.33)$$

and

$$\begin{aligned}[V^-] &= \frac{1}{2}([V]-Z_0[I]) = \frac{1}{2}([Z]-Z_0[U])[I] \\ &= ([Z]-Z_0[U])([Z]+Z_0[U])^{-1}[V^+] \\ &= ([Z']-[U])([Z']+[U])^{-1}[V^+] \end{aligned} \quad ...(5.34)$$

Relation (5.34) may be used to write

$$[S] = ([Z']-[U])([Z']+[U])^{-1} \quad ...(5.35)$$

Writing transpose of $[S]$

$$[S]_t = \{([Z']+[U])^{-1}\}_t([Z']-[U])_t \quad ...(5.36)$$

As $[Z']$ and $[U]$ are symmetrical matrices,

$$\{([Z']+[U])^{-1}\}_t = ([Z']+[U])^{-1} \quad ...(5.37)$$

and

$$([Z']-[U])_t = ([Z']-[U]) \quad ...(5.38)$$

Therefore,

$$[S] = [S]_t \quad ...(5.39)$$

This indicates that the scattering matrix $[S]$ is a symmetrical matrix.

While writing (5.28) it was assumed that all the N ports of the network are terminated in lines of equal characteristic impedance Z_0. More detailed considerations summarized below show that equality of Z_0's is not a necessary condition for symmetry of $[S]$ matrix. We define

$$\bar{V}_n = a_n + b_n = (V_n^+ + V_n^-)/\sqrt{Z_{on}} = V_n/\sqrt{Z_{on}}$$

$$\bar{I}_n = a_n - b_n = (V_n^+ - V_n^-)/\sqrt{Z_{on}} = I_n\sqrt{Z_{on}}$$

and two matrices $[\sqrt{Z_{on}}]$, $[1/\sqrt{Z_{on}}]$ are defined as two square (n by n) matrices with non-zero diagonal elements given by $\sqrt{Z_{01}}$, $\sqrt{Z_{02}}$, $\sqrt{Z_{03}}$ ..., $\sqrt{Z_{on}}$ and $1/\sqrt{Z_{01}}$, $1/\sqrt{Z_{02}}$, $1/\sqrt{Z_{03}}$ ..., $1/\sqrt{Z_{on}}$ respectively, and all non-diagonal elements as zeros.

Now equation (5.27) may be written as

$$[V] = [\sqrt{Z_{on}}][\bar{V}] = [Z][1/\sqrt{Z_{on}}][\bar{I}]$$

or
$$[\bar{V}] = [\sqrt{Z_{on}}]^{-1}[Z][1/\sqrt{Z_{on}}][\bar{I}]$$
$$= [\bar{Z}][\bar{I}]$$

where $[\bar{Z}] = [\sqrt{Z_{on}}]^{-1}[Z][1/\sqrt{Z_{on}}]$ Since $[Z]$ is a symmetrical matrix it is seen that $[\bar{Z}]$ is also a symmetrical matrix. Now following the arguments presented in eqns. (5.28) through (5.39), after replacing $[V^+], [V]$ and $[\bar{Z}']$ by $[a]$, $[b]$ and $[Z]$, respectively, it can be shown that $[S]$ is symmetrical even when all Z_0's are not equal. Thus for any lossless, linear and reciprocal network (i.e. where $[Z]$ is symmetrical) the scattering matrix is symmetrical.

5.4.3 Scattering matrix for a lossless junction

For a lossless junction the total power leaving all the N ports must equal the sum of the incident powers. The mathematical statement of this power conservation condition is

$$\sum_N (a_n a_n^* - b_n b_n^*) = 0$$

or
$$\sum_N b_n b_n^* = \sum_N a_n a_n^* \qquad ...(5.40)$$

This condition imposes a number of restrictions on the scattering matrix coefficients. These are discussed in the following. Since from equation (5.23)

$$b_n = \sum_{i=1}^{N} S_n t\, a_i \qquad ...(5.41)$$

the power conservation condition may be expressed as

$$\sum_{n=1}^{N} \left| \sum_{i=1}^{N} S_{nt} a_t \right|^2 = \sum_{n=1}^{N} a_n a_n^* \qquad ...(5.42)$$

All a_n's are independent. So if we choose all a_n's except a_i to be zero, we have

$$\sum_{n=1}^{N} |S_{ni} a_i|^2 = a_i a_i^*$$

or
$$\sum_{n=1}^{N} |S_{ni}|^2 = \sum_{n=1}^{N} S_{ni} S_n^* = 1 \qquad ...(5.43)$$

Since the index i is arbitrary, the above equation is valid for all values of i. Equation (5.43) states that for a lossless network the product of any column of the scattering matrix with the conjugate of this column equals unity.

In addition to the above constraints on the S_{nm}, another set of constraints may be derived. By selecting all a_n's except a_s and a_r to be zero ($r \neq s$), equation

(5.42) may be written as

$$\sum_{n=1}^{N} |S_{ns}a_s + S_{nr}a_r|^2$$

$$= \sum_{n=1}^{N} (S_{ns}a_s + S_{nr}a_r)(S_{ns}a_s + S_{nr}a_r)*$$

$$= |a_s|^2 + |a_r|^2 \qquad ...(5.44)$$

Expanding the left hand side and using equation (5.43) results in the concellation of a number of terms and we are left with

$$|a_s|^2 \sum_{n=1}^{N} (S_{ns}S^*_{nr}a_s a^*_r + S^*_{nz}S_{nr}a^*_s a_r] = 0 \qquad ...(5.45)$$

In view of the independent nature of a_s and a_r, we may choose, in the first instance $a_s = a_r$. This leads to

$$|a_s|^2 \sum_{n=1}^{N} (S_{ns}S^*_{nr} + S^*_{ns}S_{nr}) = 0 \qquad ...(5.46)$$

If, instead, we choose $a_s = ja_r$, with a_r real, we obtain

$$j|a_s|^2 \sum_{n=1}^{N} (S_{ns}S^*_{nr} - S^*_{ns}S_{nr}) = 0 \qquad ...(5.47)$$

Since neither a_s nor a_r is zero, both (5.46) and (5.47) are satisfied only if

$$\sum_{n=1}^{N} S_{ns}S^*_{nr} = 0 \quad s \neq r \qquad ...(5.48)$$

This equation states that *the product of any column of the scattering matrix with the complex conjugate of any other column is zero.*

Equations (5.43) and (5.48) represent two important relationships for scattering matrix coefficients for a lossless network. We will use these relations for construction of scattering matrices for certain lossless networks.

5.5 SCATTERING MATRICES FOR SOME TYPICAL NETWORKS

In this section we shall consider the structure of scattering matrices for a few simple networks.

5.5.1 Section of a uniform transmission line

Consider a section of length L of a waveguide as shown in Fig. 5.6. This section of length L can be regarded as a two port network with incident waves a_1, a_2 and outgoing waves b_1, b_2 at the ports 1 and 2, respectively. Since there are no reflections at the two ports, we have

$$b_2 = a_1 \, e^{-j\beta L} \qquad ...(5.49)$$

and

$$b_1 = a_2 \, e^{-j\beta L} \qquad ...(5.50)$$

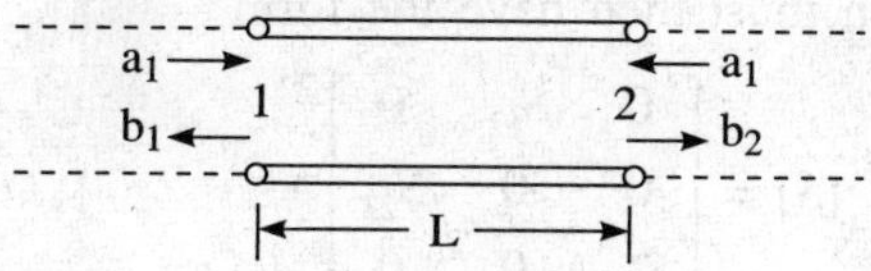

Fig. 5.6

and we may obtain the scattering matrix relation as

$$\begin{bmatrix} b_1 \\ b_2 \end{bmatrix} = \begin{bmatrix} 0 & e^{-j\beta L} \\ e^{-j\beta L} & 0 \end{bmatrix} \begin{bmatrix} a_1 \\ a_2 \end{bmatrix} \qquad ...(5.51)$$

Equation (5.51) may be used to find the modified S-matrix for any network when the reference plane for one of its ports is shifted away along the transmission line (or the waveguide) connected to that port. If an additional line length L is connected to jth port, it can be shown that each of the coefficients S_{ij} or S_{ji} (involving j) will be multiplied by the factor $e^{-j\beta L}$,while S_{jj} will be multiplied by $e^{-j2\beta L}$. The coefficients which do not involve the port j will remain unaffected.

5.5.2 A 3-port network

Properties of scattering matrix of a 3-port network may be used to show that a lossless, matched and non-reciprocal three-port microwave junction is exemplified only by a perfect three-port circulator. A perfectly matched three-port network has a scattering matrix with $S_{ii} = 0$ and the S-matrix is given by

$$[S] = \begin{bmatrix} 0 & S_{12} & S_{13} \\ S_{21} & 0 & S_{23} \\ S_{31} & S_{32} & 0 \end{bmatrix} \qquad ...(5.52)$$

For a non-reciprocal junction, the scattering matrix is no longer symmetrical that is $S_{ij} \neq S_{ji}$. However, if the junction is lossless, conservation of power still requires that the S-matrix satisfy equations (5.43) and (5.48). Applying the conditions given by (5.43) and (5.48) we have

$$S_{21}S_{21}^* + S_{31}S_{31}^* = 1 \qquad ...(5.53)$$

$$S_{12}S_{12}^* + S_{32}S_{32}^* = 1 \qquad ...(5.54)$$

$$S_{13}S_{13}^* + S_{23}S_{23}^* = 1 \qquad ...(5.55)$$

$$S_{31}S_{32}^* = S_{21}S_{23}^* = S_{12}S_{13}^* = 0 \qquad ...(5.56)$$

Let us assume that $S_{12} \neq 0$. Equation (5.56) then yields that $S_{13} = 0$. Equation (5.55) now requires that $| S_{23} | = 1$. From equation (5.56), $S_{21} = 0$. Also $| S_{31} | = 1$ (from eqn. 5.53) and $S_{32} = 0$ (from 5.56). Thus, if $S_{12} \neq 0$,

$$S_{13} = S_{21} = S_{32} = 0, \; | S_{23} | = | S_{31} | = | S_{12} | = 1$$

Consequently, there is perfect transmission from port 3 into port 2, from port 2 into port 1, and from port 1 into port 3. Also, there is no transmission in the reverse directions. The resultant scattering matrix of any matched, lossless, non-reciprocal 3-port junction must then have the form

$$[S] = \begin{bmatrix} 0 & S_{12} & 0 \\ 0 & 0 & S_{23} \\ S_{31} & 0 & 0 \end{bmatrix} \qquad ...(5.57)$$

Alternatively, it may have the following form

$$[S] = \begin{bmatrix} 0 & 0 & S_{13} \\ S_{21} & 0 & 0 \\ 0 & S_{32} & 0 \end{bmatrix} \qquad ...(5.58)$$

If the locations of the terminal planes in the three input lines are properly chosen, the phase angles of S_{13}, S_{21} and S_{32} can be made zero and then $S_{13} = S_{21} = S_{32} = 1$.

If the above analysis were applied to a lossless, perfectly matched, reciprocal 3-port junction, reciprocity would require that

$$S_{13} = S_{21} = S_{32} = 0$$

Since S_{31}, S_{12}, S_{23} are zero, the resultant matrix would then vanish. It can thus be concluded that it is not possible to construct a perfectly matched, lossless, reciprocal 3-port junction. At least one of the reflection coefficients must be different from zero in the reciprocal case.

5.5.3 A 3-dB power divider

For a power divider, the input port (numbered 1) is matched and the incident power is divided equally in two output ports 2 and 3. This can be expressed in terms of scattering matrix coefficients as

$$S_{11} = 0 \qquad ...(5.59)$$

$$S_{21} = 1/\sqrt{2} \qquad ...(5.60)$$

$$S_{31} = 1/\sqrt{2} \qquad ...(5.61)$$

Equal power division implies only $| S_{21} | = | S_{31} | = 1/\sqrt{2}$. However, by proper selection of reference planes we could justify (5.60) and (5.61). With this information, the S-matrix may be constructed as follows:

$$[S] = \begin{bmatrix} 0 & 1/\sqrt{2} & 1/\sqrt{2} \\ 1/\sqrt{2} & S_{22} & S_{23} \\ 1/\sqrt{2} & S_{23} & S_{33} \end{bmatrix} \qquad ...(5.62)$$

Using properties of scattering matrices for lossless network given by equation (5.43) we have

$$\frac{1}{2} + S_{22}S_{22}^* + S_{23}S_{23}^* = 1 \quad ...(5.63)$$

$$\frac{1}{2} + S_{23}S_{23}^* + S_{33}S_{33}^* = 1 \quad ...(5.64)$$

These two relations lead to

$$|S_{22}|^2 = |S_{33}|^2 \quad ...(5.65)$$

Also by using equation (5.48) we have,

$$1/\sqrt{2}\, S_{22}^* + 1/\sqrt{2}\, S_{23}^* = 0 \quad ...(5.66)$$

which yields

$$S_{22} = -S_{23} \quad ...(5.67)$$

Substituting this in (5.63) we have

$$\frac{1}{2} + S_{22}S_{22}^* + S_{22}S_{22}^* = 1$$

or

$$|S_{22}|^2 = 1/4 \quad ...(5.68)$$

So,

$$|S_{22}| = |S_{23}| = |S_{33}| = 1/2 \quad ...(5.69)$$

Relation (5.69) clearly indicates that ports 2 and 3 are not matched. Voltage reflection coefficient at each one of these ports will be 0.5.

The three network examples described above demonstrate that the properties of scattering matrices may be used to study the characteristics of microwave networks.

SELECT BIBLIOGRAPHY

R.E. Collin, *Foundations for Microwave Engineering,* McGraw-Hill Book Company 1966, Chapter 4, on Circuit Theory for Waveguiding Systems.

S. Ramo, J.R. Whinnery and T. Van Duzer, *'Fields and Waves in Communication Electronics'*, Section 5.02 to 5.12 on The Formulation of a Circuit Concept Consistent with Maxwell's Equations.

H.A. Atwater, *Introduction to Microwave Theory,* McGraw-Hill Book Company, Inc. 1962, Chapter 4, on Discontinuities and Impedances in Waveguides.

Hewlett Packard, "*S*-parameters, Circuit Analysis and Design", Application Note 95, September, 1968 (A collection of articles describing S-parameters, circuit design and analysis).

PROBLEMS

5.1 A standard *X*-band waveguide with inside dimensions 0.9 inch by 0.4 inch is connected to another waveguide which is 0.8 inch by 0.3 inch. The latter guide is terminated in a matched load. Calculate VSWR that would be caused by this discontinuity at 10 GHz. (Reactances associated with the geometrical discontinuity may be ignored.)

5.2 A lossless transmission line of characteristic impedance Z and length L is terminated on both ends with transmission lines of characteristic impedance Z_0. Derive an expression for the scattering matrix coefficient S_{11}. What is the value of S_{11} when L is quarter wavelength?

5.3 Lossless transmission line of length L (characteristic impedance Z_0 and phase constant β) is connected at the input terminals of a 2-port network with scattering coefficients S_{11}, S_{12}, S_{21} and S_{22}. Find the scattering matrix coefficients for the composite network.

5.4 Show that for a symmetrical 3-port reciprocal waveguide junction S_{nn} = 1/3. By attaching a double stub tuner at port 1, it is possible to make $S_{11} = 0$ What will be the values of the other coefficients of the scattering matrix for the combination? Assume that the terminal planes are selected such as to make all scattering coefficients real.

5.5 Calculate *S*-parameters for the waveguide junction described in Problem 5.1.

5.6 An admittance $Y(= G + jB)$ mhos is connected across a uniform transmission line of characteristic admittance Y_0 as shown in Fig. 5.7. Derive the *S*-matrix for this 2-port. Does this *S*-matrix satisfy the relation 5.43? Also, examine this for $G = 0$ and explain.

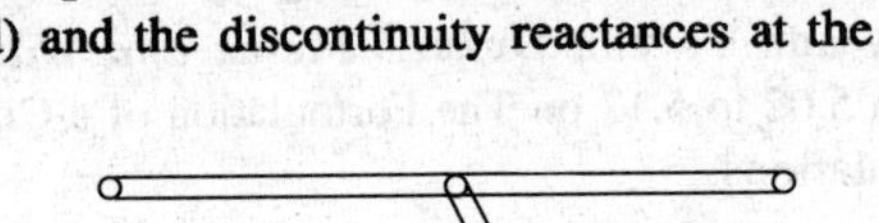

Fig. 5.7

5.7 Consider a *T*-junction shown in Fig. 5.8. Derive the scattering matrix for this 3-port junction. Reference planes may be chosen near the junction (i.e. line lengths need not be considered) and the discontinuity reactances at the junction may be ignored.

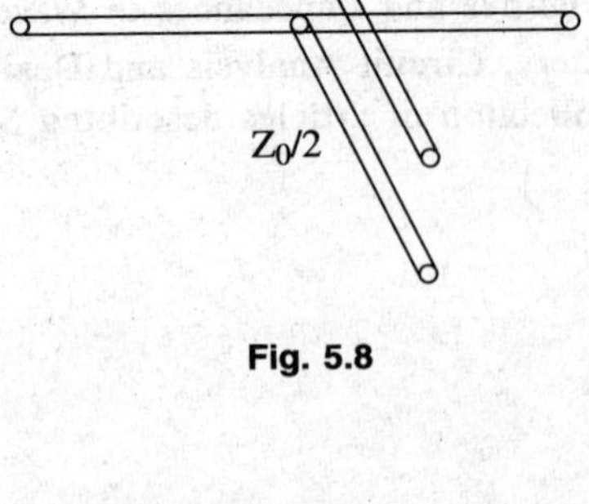

Fig. 5.8

5.8 Figure 5.9 shows a microstrip line of impedance Z_1 connected to another line of higher impedance Z_2. Write the S-matrix for the junction. If the discontinuity reactance is represented by a series inductance (jX) as shown in the figure, derive the modified S-matrix taking jX into account.

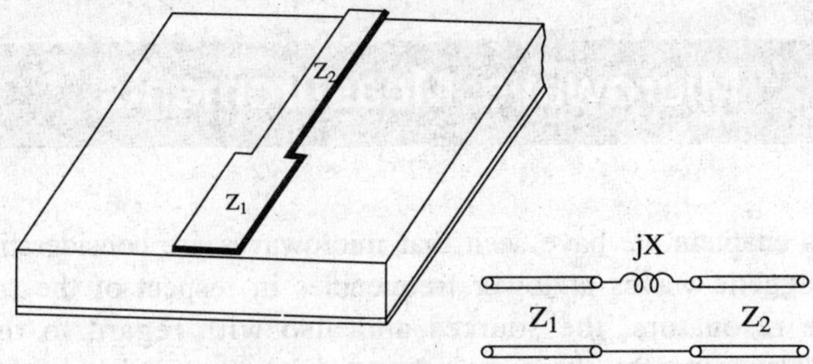

Fig. 5.9

5.9 A two-port network with S-parameters S_{11}, S_{12}, S_{21} and S_{22} is terminated with an impedance Z_L at port 2. Calculate the input reflection coefficient at port 1.

5.10 Can the measurement of input reflection coefficient in Problem 5.9 be used for measurement of S-parameters by varying Z_L. How many known values of Z_L's will be required and which scattering parameters can be measured in this manner?

CHAPTER 6

Microwave Measurements

In the previous chapters we have seen that microwaves are considerably different from electromagnetic waves at lower frequencies in respect of the transmission structures, the resonators, the sources and also with regard to the network representation. Consequently, the measurement techniques, and even the quantities that can be measured are different for the microwave frequency range.

As pointed out in Chapter 5, commonly used definitions of voltage and current become vague at microwave frequencies. Therefore, power and impedance become the principal measurable quantities. Measurement of microwave power is based on either the detection of microwaves by a non-linear resistor or by the measurement of heat produced when a certain amount of microwave power is dissipated in a resistive load. Detection of microwave signals is generally used for relative power measurements. This is generally the case when two power levels are to be compared, such as in the measurement of standing wave ratio in waveguides, in which case only the ratio of maximum to minimum voltage is desired. On the other hand, the methods based on thermal effects are used when absolute power levels (in watts) are required.

Measurement of impedance at microwave frequencies is accomplished by measuring the reflection coefficient caused by the given unknown impedance when it is connected to the end of a transmission line (or waveguide) of known characteristic impedance. As discussed in Chapter 2 for the case of transmission lines, the reflection coefficient is obtained by measuring the voltage standing wave ratio along the line and the position of the first minimum from the load. Various methods of impedance measurement are discussed in this chapter.

For representation of microwave networks, it becomes necessary to measure S-parameters discussed in Chapter 5. These are derived from the measurement of input and output power-levels (and phase components) at various ports.

Since various characteristics of microwave components and networks are functions of frequency, an accurate measurement of operating frequency becomes desirable. Methods of frequency measurement are also described in this chapter.

6.1 DETECTION OF MICROWAVES

'Detection' of microwaves is carried out for relative measurement of microwave power. In order to indicate relative power levels, it is necessary to 'detect' or rectify the microwave signals and obtain a proportionate DC signal. The principle of rectification is similar to that used at lover frequencies.

6.1.1 Detector diodes

Diodes used at lower frequencies, both the vacuum tubes and the semiconductor diodes, are not as such useful for microwave frequencies. Vacuum tube diodes suffer from 'transit time effects' described in Chapter 3. Ordinary semiconductor diodes have large junction capacitance which make them unsuitable for use at microwave frequencies. The diodes that can be used for detection of microwaves are specially designed point contact diodes or metal-semiconductor Schottky barrier diodes.

Point contact diodes consist of a thin tungsten 'whisker' touching a silicon chip. A typical construction is shown in Fig. 6.1. The fabrication starts with a thin wafer of poly-crystalline *p*-type silicon (normally doped with aluminium). The wafer surface is cleaned chemically and polished flat. A thin oxide layer is grown by heating the wafer to about 1200°C in steam or oxygen. During oxidation some *p*-type impurities diffuse out (into oxide) and a high resistivity layer (shown by dotted line in Fig. 6.1) is formed. Oxide layer is then removed and the wafer divided into small chips.

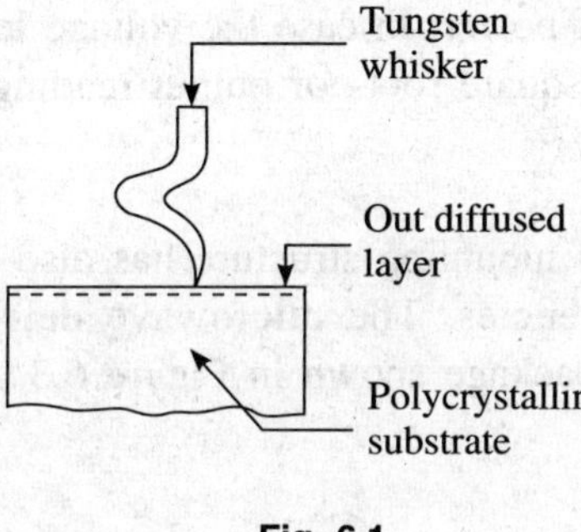

Fig. 6.1

The tungsten whisker is etched electrolytically to obtain a tip diameter of the order of 12 μm. After assembly a process of 'adjusting' is necessary for obtaining the required diode characteristics. Adjusting is the process of increasing the tungsten whisker force by exciting mechanical resonances. This is usually carried out by tapping the package with a small hammer. Even today a large percentage of microwave diodes are made by this technique.

Alternative to point contact diode is a metal-semiconductor Schottky barrier diode. Typical construction of this type of diode is shown in Figure 6.2. The fabrication is carried out by 'planar technology'. It consists of a set of processes which are also used for production of transistors and integrated circuits (mentioned in Chapter 4). A thin slice of silicon (n^+-type) is cleaned and polished flat. This is called the substrate.

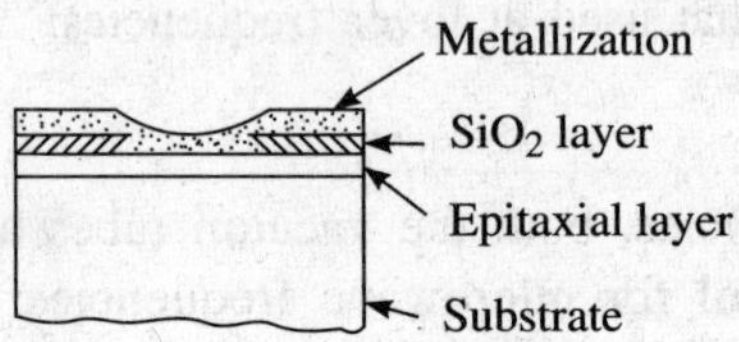

Fig. 6.2

An n-type active layer about 2 micron thick is grown epitaxially over the substrate. A thin SiO_2 layer is grown thermally over the active layer. Windows are etched in the oxide layer to open out areas for metal-semiconductor junction. The metal is deposited by vacuum evaporation. Excess metal over the surrounding areas is etched out by photo-etching process.

Schottky barrier diodes have several advantages over point contact diodes. These advantages are higher burn out rating, lower $1/f$ noise and better reliability. Most of these features result from the use of superior and advanced technology. For this reason Schottky barrier diodes are gradually replacing point contact diodes.

Both the point contact and the Schottky barrier diodes exhibit a 'square law' response under normal operating conditions. Output current is proportional to the square of the electric field value at the location of the diode. Thus the output current is a direct measure of power. In case the voltage levels are to be compared (as in VSWR measurement) square roots of output readings may be taken.

6.1.2 Detector mounts

Not only the detector, but its mounting structure has also to be designed carefully for use at microwave frequencies. The microwave detector diodes are usually available in the type of the package shown in Figure 6.3.

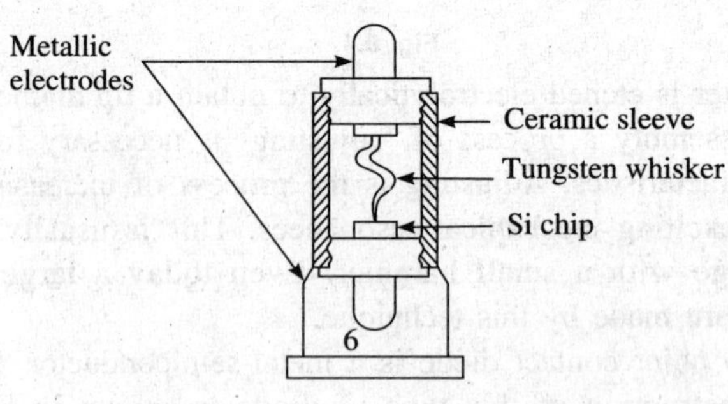

Fig. 6.3

The package mounting in the circuit is designed carefully so that the parasitic reactances introduced are minimum.

The detector diode may be mounted in a coaxial line (Fig. 6.4) or a waveguide (Fig. 6.5) depending upon the type of the measurement system that is being used. In both the cases, the diode is mounted across the line (or waveguide) and the other end of the line (or waveguide) is shorted.

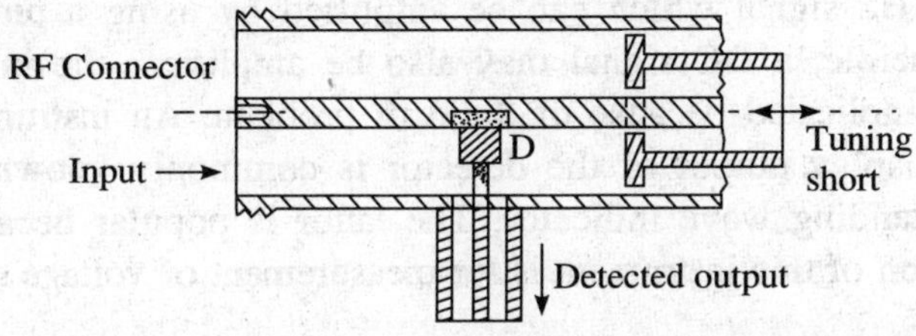

Fig. 6.4

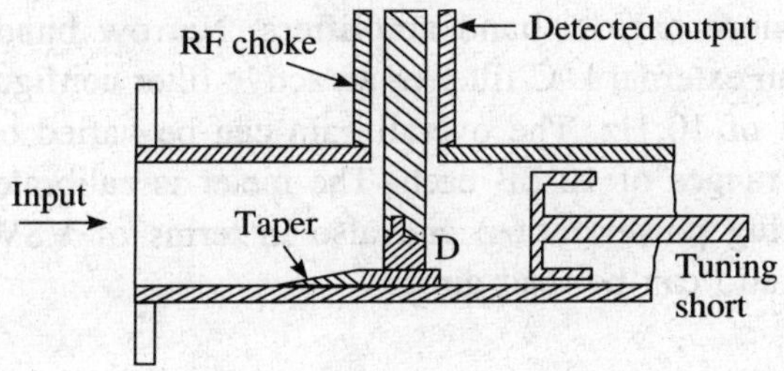

Fig. 6.5

The distance between the shorted end and the diode can be varied by moving a tuning plunger. The tuning arrangement helps the user to readjust the distance of the diode from the short circuit whenever the signal frequency is changed. In this way, detector performance may be optimized at the desired frequency. The detected output is normally available at a coaxial (BNC) connector.

The performance of microwave detectors is expressed in terms of the current sensitivity which is defined as follows:

$$\beta = \frac{\Delta i}{P} \qquad \text{...(6.1)}$$

where Δi is the increase in short-circuit current resulting from an available input power P. Sensitivity β is a function of bias level and is usually in the range of about 1 to 15 μA/μW. The highest values are usually attained when the bias level is between 10 and 100 μA. However, because of circuit simplicity, most of the detectors are operated without any applied bias.

6.1.3 Detector output indicator

The detector diode in the mounting arrangement described above will detect a microwave signal and give a direct current proportional to the microwave power at the input. The output current is typically few microamperes or tens of microamperes and can be read by connecting a DC microammeter. This arrangement, however, gives a poor sensitivity.

For increasing the sensitivity of the detector measurements it is necessary to modulate the microwave power at the source itself by a low frequency signal. Typically 1 KHz modulation frequency is used. In such a case the output of the detector is a 1 KHz signal which can be amplified by using a tuned amplifier. Although, in principle, a DC signal may also be amplified, the instrumentation becomes more complicated because of the drift problem. An instrument designed to amplify and display output of the detector is commonly known as 'detector amplifier' or a standing wave indicator. The latter is popular because the most common application of this instrument is for measurement of voltage standing wave ratio.

Typically a 'detector amplifier' or 'standing wave indicator' can be described by a block diagram shown in Fig. 6.6. There are several stages of amplification including one or more narrow band amplifiers. Narrow band amplification is achieved either by an external L-C filter or an active filter configuration. Bandwidth here is of the order of 10 Hz. The overall gain can be varied over a range of 60 dB in six different ranges of 10 dB each. The meter is calibrated to read directly in dBs (for measuring power ratios) and also in terms of VSWR values so that the standing wave ratio can be read directly.

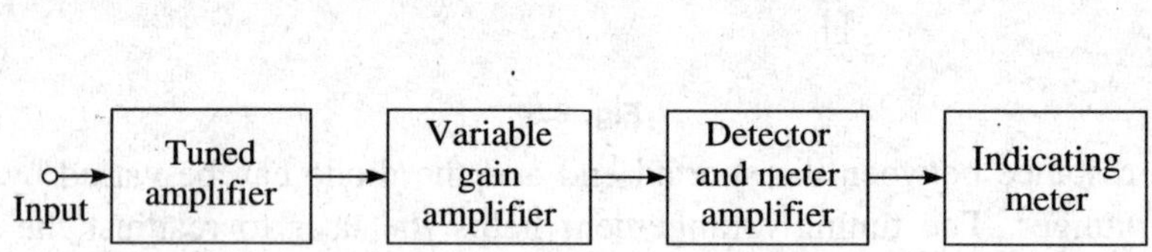

Fig. 6.6

6.2 MICROWAVE POWER MEASUREMENT

6.2.1 Bolometers

The detector-amplifier method described in the previous section is not suitable for an absolute measurement of the microwave power (in watts). Microwave power meters used for this purpose are based on the conversion of microwave power into thermal energy. These instruments employ sensing devices called bolometers. Microwave power incident on the bolometer raises its temperature. The change in the temperature causes variation in the DC resistance of the bolometer. There are two general classifications of bolometers.

(*a*) *Positive temperature coefficient type*: These are conductors, of which barretter is a typical example. Barretter consists of a very fine platinum wire mounted in a holder which is hermetically sealed and permits easy measurement of resistance changes. Time constant of the barretters is of the order of 100 μsec.

(b) ***Negative temperature coefficient type:*** These are semiconductors. The most popular type is thermistor. Thermistor is constructed in the form of a small bead of semiconducting material suspended between two fine wires. This tiny bead, about 0.04 cm in diameter, is composed of a mixture of the oxides of manganese, cobalt, nickel and copper. It can be mounted directly in a waveguide. Thermistors have longer time constants (of the order of a second) than barretters. The thermistor is basically more sensitive than the barretter but it is also much more sensitive to the changes in the ambient temperature.

Barretter mounts are very similar to detector diode mounts. Thermistors, however, in most cases require a special arrangement to take into account the ambient temperature variations.

6.2.2 Microwave power meters

These instruments are designed to process the output of 'bolometer' and to represent the power level on a calibrated scale.

One of the simplest methods is to place the bolometer in one arm of a Wheatstone bridge as shown in Fig 6.7. The bridge is energized by a regulated DC supply whose amplitude may be adjusted with R_1. Since R_4 is a thermistor, its resistance may be controlled by the heating caused by the current through it and is thus adjusted equal to R_5, bringing the bridge into balance and causing the meter to read zero. Microwave power is then applied to the thermistor R_4 which is mounted in a waveguide or a coaxial line. Heating effect causes the thermistor resistance to decrease and unbalances the bridge in proportion to the power applied. This unbalance current is indicated by the meter, which is calibrated directly in milliwatts.

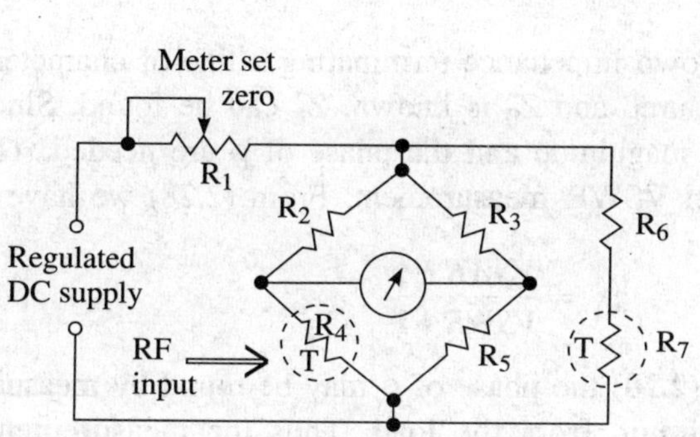

Fig. 6.7

The unbalanced bridge technique is seldom used in commercial microwave power meters. Circuit modifications commonly used include self-balancing bridge and a compensation against the variations in ambient temperature. A self-balancing bridge operates on the principle of power substitution. Initially some DC plus some audio-frequency power is applied to the bolometer and the bridge is balanced. When the microwave power (to be measured) is incident on the bolometer, the balance of the bridge is disturbed. This unbalance of the bridge is sensed by an electronic circuit which is so arranged that an equivalent amount of audio-frequency power is removed from the bolometer and the balance restored. This reduction in audio-frequency power is measured with a VTVM and indicates the microwave power under measurement. In a temperature compensated power meter, another thermistor (R_7 in Fig. 6.7) is placed in close thermal proximity of the thermistor on which the microwave power is incident. This additional thermistor R_7 is isolated from the microwave power but attains a temperature close to that of the microwave thermistor R_4. Change in resistance of R_7 (due to ambient temperature variations) compensates for corresponding change in R_4 by controlling the DC power that is applied to the bridge.

6.3 IMPEDANCE MEASUREMENT

6.3.1 Basic principles

The simplest method for the measurement of impedance at microwave frequencies is as follows. The unknown impedance is connected at the end of a slotted coaxial line (or a slotted waveguide). Microwave power is fed from the other end of the coaxial line (or waveguide). Unknown impedance reflects a part of this power. This reflection coefficient is measured by probing the standing wave fields in the slotted line (or waveguide) by a suitable arrangement. As discussed in Chapter 2 (see equation 2.26) the reflection coefficient is given by

$$\rho = \frac{Z_L - Z_0}{Z_L + Z_0} \qquad ...(6.2)$$

where Z_L is the unknown impedance terminating a line of characteristic impedance Z_0. Thus if ρ is measured and Z_0 is known, Z_L can be found. Since in general Z_L is complex, both the magnitude and the phase of ρ are needed. The magnitude of ρ may be found from VSWR measurement. From (2.28) we have

$$|\rho| = \frac{VSWR - 1}{VSWR + 1} \qquad ...(6.3)$$

Also, by using (2.29) the phase of ρ may be found by measuring the distance of first voltage minimum from the load. Thus the measurement of impedance involves the measurements of the VSWR and the distance of voltage minimum from the load. These measurements may be carried out by using a slotted line and probe arrangement.

6.3.2 Slotted line and probe

Slotted line is a fundamental tool for microwave measurements. It consists of a section of waveguide or coaxial line (for measurements in waveguide and coaxial line systems, respectively) with a longitudinal slot. This slot is roughly 1 mm wide and allows an electric field probe to enter the waveguide (or the coaxial line) for measurement of the relative magnitude of field at the location of the probe. The slot is located suitably in the wall of the waveguide such that the disturbance to the wall currents is minimum. For a rectangular waveguide operated in the dominant mode this location is in the middle of the broad wall. The location of the slot and the direction of the wall currents in waveguide and coaxial system is shown in Figs. 6.8 (a) and (b), respectively. Slotted section is normally mounted in a 'carriage' which also supports the probe moving inside the slot.

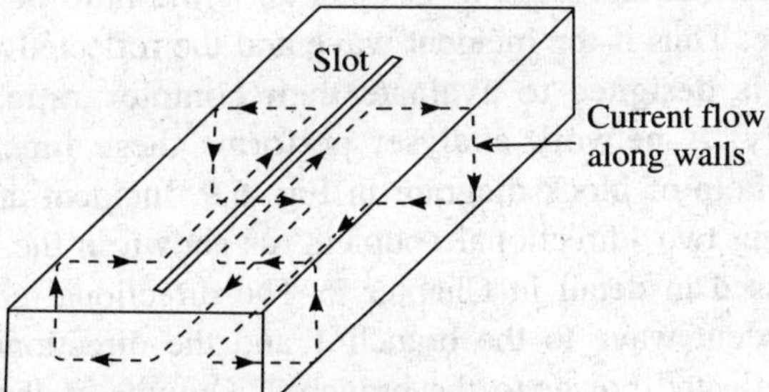

Fig. 6.8 (a)

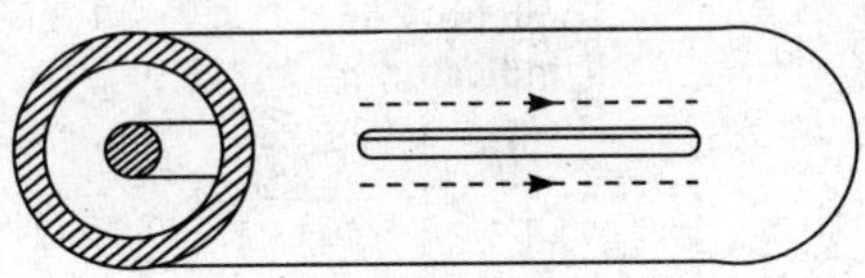

Fig. 6.8(b)

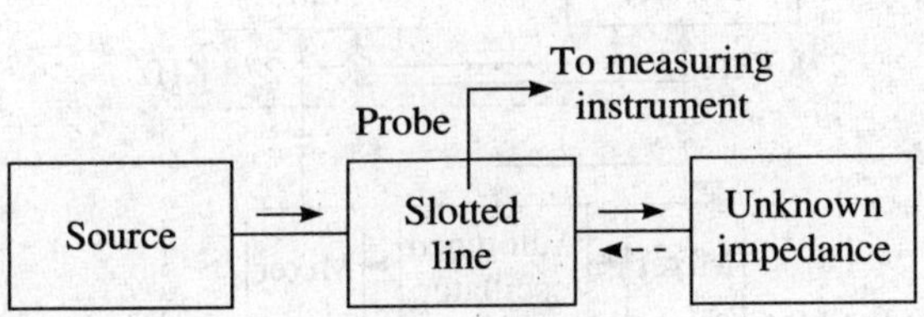

Fig. 6.8(c)

Probe is a thin conducting wire which passes through the slot in the slotted line and couples to the fields in the waveguide. The amount of insertion of the probe in the line needs careful adjustment. A small insertion results in a poor sensitivity and makes it difficult to locate the minima of the standing wave pattern. Too deep an insertion, on the other hand, disturbs the field in the line and may in fact reflect some of incident input power and thereby lead to inaccurate results. The output of the probe is connected to the detector in a mount (coaxial) similar to that shown in Fig. 6.3.

The measurement of impedance consists of mounting of the unknown impedance at the end of a slotted line as shown in Fig. 6.8 (c) and noting VSWR and d_{min} by moving the probe and observing the output on a VSWR indicator. Knowing VSWR, and d_{min}, a graphical construction on Smith chart may be used to calculate Z_L. This procedure has been already discussed in Chapter 2.

6.3.3 Impedance measurement by network analyser

The method of impedance measurement discussed above is fairly time-consuming and does not lend itself to automation. Use of a network analyser is more convenient when rapid measurements over a broad frequency range are required or when variations of impedance with respect to some circuit parameter is to be monitored.

This method is based on direct measurement of complex reflection coefficient. Reflection coefficient (as discussed in Chapter 2) is the ratio of the reflected wave to the incident wave. Thus if the incident wave and the reflected wave are separated and a mechanism is designed to evaluate their complex ratio, we can measure impedances directly. A network analyser performs these functions and may be described with the help of block diagram in Fig. 6.9. Incident and reflected waves are sampled by using two 'directional couplers' as shown in the figure. Directional couplers are discussed in detail in Chapter 7. The directional coupler 1 couples a fraction of the incident wave to the branch 1 and the directional coupler 2 feeds a fraction of the reflected power to the branch 2. Outputs of these two directional

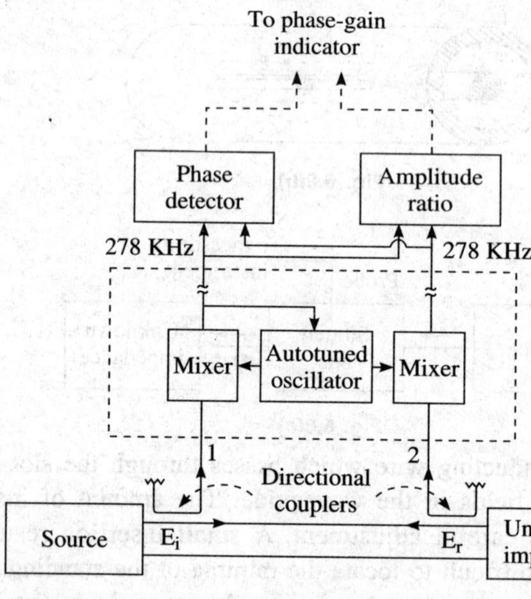

Fig. 6.9

couplers are fed to a harmonic frequency converter which translates the frequency from the microwave range to a fixed frequency of 278 KHz. An auto-tuning local oscillator and two identical mixers are used for this purpose. The frequency conversion is usually carried out in two steps. The second set of mixers is not shown in the diagram. The two outputs at 278 KHz preserve the relative amplitude and the phase relationship of the incident and the reflected waves at microwave frequency. Phase comparison and amplitude comparison is now carried out at 278 KHz frequency by suitably designed low frequency circuits. The phase and amplitude information may now be given to a CRO for direct display of the impedance on a Smith chart overlay placed over the CRO screen.

A network analyser of the type discussed above is an extremely versatile instrument and may be used for several other measurements including the measurement of scattering parameters discussed in the next section.

6.4 MEASUREMENT OF SCATTERING PARAMETERS

As discussed in Chapter 5, scattering parameters are defined as the ratios of the outgoing waves to the incident waves. Thus the parameter s_{nm}, is the ratio of b_n

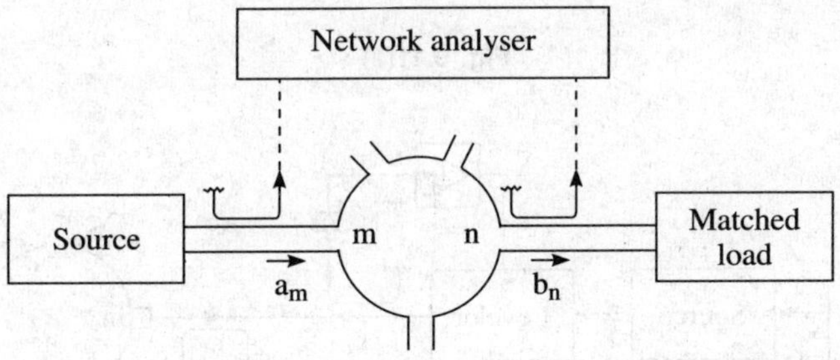

Fig. 6.10

to a_m (Fig. 6.10). Measurement of S_{nm}, involves the following steps:

(i) Connect a source to the port *m* and a matched load to the port *n*. These two ports are connected to a network analyser *via* two directional couplers as shown in Fig. 6.10.

(ii) Terminate all other ports in matched loads so that all other *a*'s (the amplitudes of the incident waves) are zero.

(iii) The network analyser indicates the complex value of S_{nm} in terms of amplitude and phase.

In many cases only the amplitude of the scattering matrix coefficient S_{nm} is needed. The measurement set-up can then be simplified considerably. In this case, signals coupled to the two directional couplers are measured by a power meter or a 'detector amplifier'. Ratio of the two amplitudes yields the value of $|S_{nm}|$. Only relative amplitude levels are required.

The above method can be modified to obtain swept-frequency measurement of S-parameters over the desired frequency range. In swept frequency measurements the source frequency is varied continuously and $|S_{nm}|$ is obtained as a function of frequency either on a CRO screen or on an $X - Y$ recorder. Experimental set-up used for swept-frequency measurements is shown in Fig. 6.11 (a). The corresponding set-up for measurement of $|S_{mm}|$ is shown in Fig. 6.11 (b). Basic requirement is a 'sweep oscillator', i.e. a source whose output frequency can be

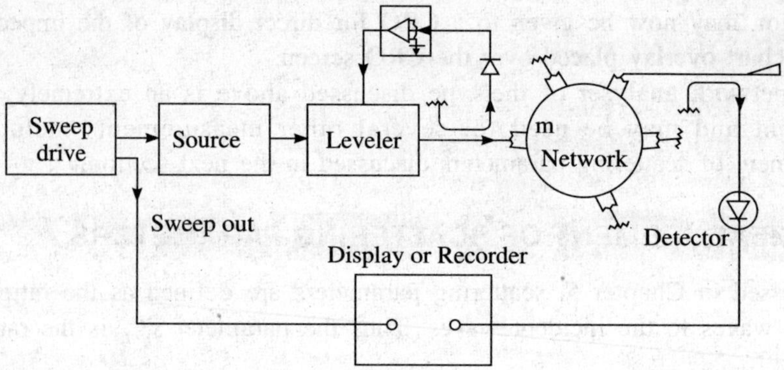

Fig. 6.11(a)

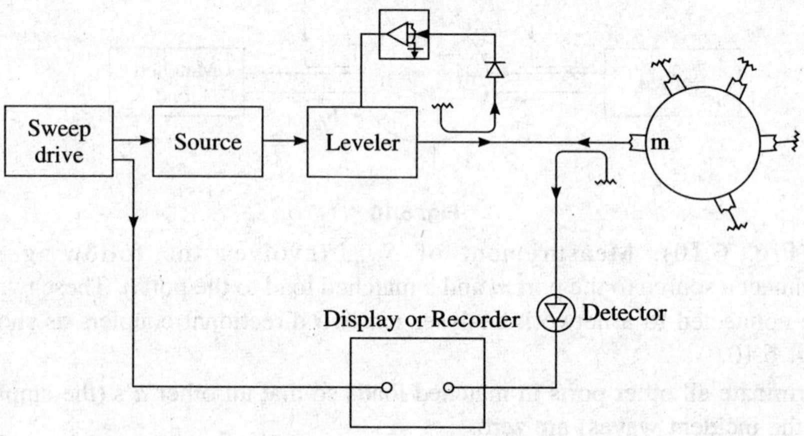

Fig. 6.11(b)

varied by a suitable voltage or current drive. Output of the sweep oscillator is passed through an automatic levelling circuit (ALC) which maintains the power output at a constant level. ALC consists of a variable attenuator (using PIN diodes discussed later in Chapter 9), a directional coupler to sample the power output and a feedback loop consisting of a detector and a DC amplifier. If at a certain frequency the output of the sweep oscillator goes down, the corresponding detector current is reduced and the feedback reduces the attenuation in the 'leveler' to maintain the output of the leveler at the desired value. Thus ALC ensures that the incident power or a_m remains constant. The scattering parameter S_{nm} is now proportional to the output at nth port. By suitable calibration $| S_{nm} |$ can be observed and recorded directly.

Swept frequency measurements may be carried out with network analyser also. In this case both the magnitude and the phase of S_{nm} are obtained as functions of frequency.

6.5 FREQUENCY MEASUREMENT

Microwave frequency can be measured by a number of different mechanical and electronic techniques. The mechanical devices commonly use circuit elements such as slotted lines and resonant cavities, both of which depend on a precise measurement or calibration of physical dimensions for their operation and accuracy. The electronic devices are primarily frequency counters and high frequency heterodyne systems which compare harmonics of a known lower frequency with the unknown microwave frequency. Although electronic frequency measuring systems are usually more complex and expensive than the mechanical methods of measurement, the electronic systems are capable of considerably higher accuracy.

6.5.1 Use of slotted line

This technique depends on the fact that standing waves set up in a transmission line or a waveguide produce minima every half wavelength apart. If these minima are detected and the distance between them is measured, the guide wavelength may be determined. Frequency may be calculated from the value of guide wavelength and the physical dimensions of the waveguide. Accuracies obtained with this technique are usually limited to approximately 1 per cent, because the guide wavelength depends critically on the waveguide dimensions. Frequency measurement in coaxial systems is not dependent upon the physical dimensions (cross-section) of the slotted coaxial line.

6.5.2 Use of resonant cavities

This constitutes the most commonly used type of microwave frequency meters (also called wave-meters). The key element is a cylindrical or coaxial resonant cavity. A simple sketch of commonly used frequency meter is shown in Fig. 6.12. The size of the cylindrical cavity can be varied by an adjustable plunger which can be moved by a calibrated dial knob assembly. The range of the meter depends upon the tuning range of the cavity. The design of the cavity is such that for a given position of the plunger, the cavity is resonant only at one frequency in the specified range. The Q factor of the cavity is made very high, often as high as 5,000.

The cavity is coupled to the waveguide through an iris in the narrow wall of the waveguide as shown in Fig. 6.12. If the frequency of the wave passing through the waveguide is different from the resonance frequency of the cavity, the transmission through the waveguide is not affected. On the other hand, if these two frequencies coincide a resonant field is set up inside the cavity, and because of the power loss associated with the cavity the wave passing through the waveguide is attenuated (roughly by 1 to 3 dB). If there is an indicating instrument (standing wave indicator or the power meter) connected such that the frequency meter is in between the source and the indicator, the indicating instrument will show a dip. Since the movement of the cavity plunger is calibrated, the signal frequency can be read.

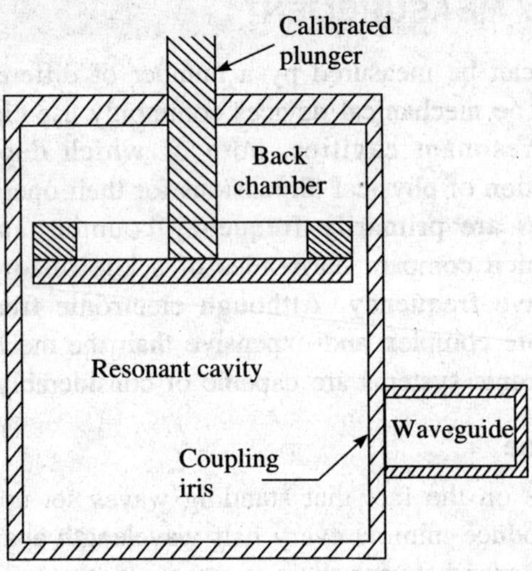

Fig. 6.12

Accuracy of 0.1 per cent or better can be obtained with this instrument.

The resonant frequency of a cavity frequency meter is determined primarily by its physical configuration and the dielectric constant of the medium inside the cavity. Consequently, these frequency meters are commonly affected by temperature changes (which cause differential expansions in the cavity) and humidity (which causes a slight change of the dielectric constant of the air inside the cavity). These effects can be minimized by use of hermetically sealed cavity and special temperature compensating materials.

6.5.3 Electronic techniques for frequency measurement

These techniques are based on the comparison of the unknown frequency with a harmonic of the known standard frequency. Instruments that use null-beating technique for this comparison are known as transfer oscillators. There are also automatic transfer oscillators which make direct display of microwave frequencies possible. Basic principles of operation of these instruments are discussed below.

TRANSFER OSCILLATORS

A typical block diagram for a transfer oscillator is shown in Fig. 6.13. Signal from a stable frequency source is amplified and fed to a harmonic generator which provides a comb of frequencies in the desired microwave range. This harmonic output is mixed with the unknown signal frequency.

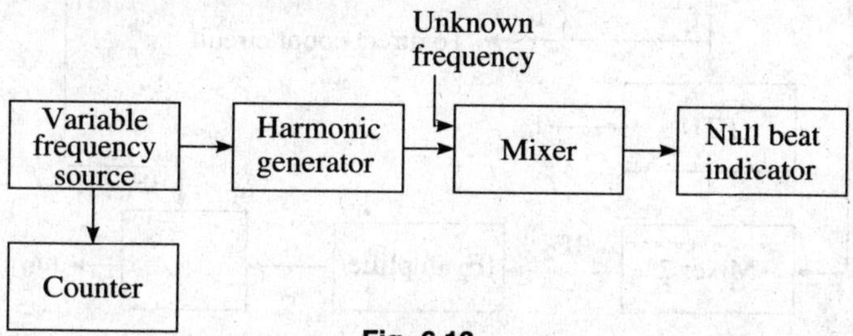

Fig. 6.13

If the frequency of the stable frequency source is varied and the mixer output connected through a detector to an indicator such as an oscilloscope, beat frequency output can be observed. Frequency of the stable frequency source is measured for the null beat condition. The unknown frequency is an integral multiple of this value. The value of the integral multiplier can be obtained by noting the next higher frequency which produces null-beat. If two frequencies are f_1 and f_2, the unknown frequency f may be found by the following equations:

$$nf_1 = f \quad \text{...(6.4a)}$$

$$(n-1)f_2 = f \quad \text{...(6.4b)}$$

Eliminating n we may write

$$f = f_1 f_2/(f_2 - f_1) \quad \text{...(6.5)}$$

Since f_1 and f_2 can be measured very accurately this method provides very precise frequency measurements. In fact, this method is often used to calibrate resonant cavity type wavemeters described earlier.

DIRECT READING MICROWAVE COUNTERS

The basic idea is similar to that in transfer oscillators. Variation in the frequency of low frequency source and search for the 'null-beat' are now made automatic by using a phaselock loop arrangement.

The block diagram of such an automatic transfer oscillator is shown in Fig. 6.14. The input signal is divided into two equal signals by a resistive power divider. These two parts of the signal are fed to two wide band harmonic mixers. The harmonic mixer 1 is used in the input phase lock loop and the mixer 2 is used to determine the harmonic number. The frequency f_1 of the input phase lock loop is also fed to the direct count circuits.

The input phase-lock loop consists of a voltage controlled oscillator (VCO), a wide band harmonic mixer, an *IF* amplifier, a phase detector, and a gain control block marked KG(s). The VCO searches over its range until an *IF* signal equal to

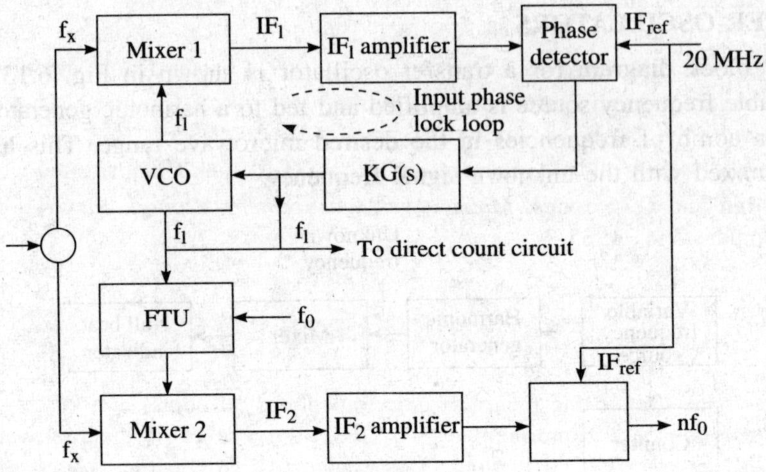

Fig. 6.14

the reference frequency of 20 MHz is found. Phase lock occurs when the phase detector output sets the VCO frequency f_1 such that:

$$f_x = nf_1 - IF_1 \quad ...(6.6)$$

where $IF_1 = IF_{ref} = 20$ MHz at the phase lock and f_x is the unknown frequency to be measured.

Next, the VCO frequency f_1 is translated to a frequency f_2 so that

$$f_2 = f_1 \pm f_0 \quad ...(6.7)$$

where the offset frequency f_0 is 20 KHz.

This is done in a frequency translation unit (FTU). The frequency f_2 drives the second sampler and produces a second *IF* output IF_2 as follows:

$$\begin{aligned} IF_2 &= nf_2 - f_x \\ &= n(f_1 \pm f_0) - (nf_1 - 20 \text{ MHz}) \\ &= \pm nf_0 + 20 \text{ MHz} \end{aligned}$$

By mixing IF_2 with the 20 MHz reference frequency and rejecting 20 MHz and higher frequencies, the harmonic frequency, nf_0, is obtained. Counting the number of zero crossing for the period of the offset frequency $f_0 = 20$ KHz then determines the harmonic number *n* of the phase lock loop. The input frequency is then calculated and displayed by presetting into the counter the 20 MHz reference frequency, measuring the VCO frequency f_1 and extending the gate time according to the harmonic number *n*.

The instruments based on the above principle can provide a direct readout of frequency over a wide range extending, for example, from 10 Hz to 18 GHz.

SELECT BIBLIOGRAPHY

S.F. Adam, *Microwave Theory and Applications*, Prentice-Hall Inc., 1969.

Hewlett-Packard Company, Engineering Staff, Microwave Division, *Microwave Theory and Measurements,* Prentice-Hall Inc., 1962.

A.J. Baden Fuller, *Microwaves,* Pergamon Press, 1969, Chapter 12 on Measurements.

R.F. Schneider, 'A High Performance Automatic Microwave Counter', *Hewlett Packard Journal,* Vol. 24, No. 8, April 1973, pp. 1–9.

E.L. Ginzton, *Microwave Measurements,* McGraw-Hill Book Company Inc., 1957.

R. Kellejian and C.L. Jones, *Microwave Measurement Manual,* McGraw-Hill Book Company Inc., 1965.

PROBLEMS

6.1 A coaxial slotted line (characteristic impedance 50 ohms) is used for measuring impedance of a component at 5.0 GHz. True value of the impedance being measured is $(20 + j\,10)$ ohms. If the accuracy with which the position of the slotted line probe can be read is ± 0.3 mm, calculate the accuracy obtained in the impedance measurement. Neglect other sources of error that may exist.

6.2 A coaxial slotted line (50 ohms) arrangement is used to find impedance of a load of 40 ohms. If the uncertainty in the VSWR measurement is ± 0.25 dB, calculate the error that will be introduced in the impedance value. Ignore the inaccuracy in the probe position and any other source of error.

6.3 The impedance measurement discussed in Problem 6.1 above is now carried out by using a network analyser set-up shown in Fig. 6.9. However, the two directional couplers do not have identical characteristics. If their coupling factors (ratio of the coupled power to the power in the main line) are 20 dB ±0.5 dB in each case, calculate the maximum error that could be introduced in the impedance measurement.

6.4 A resonant cavity type frequency meter (shown in Fig. 6.12) is made of brass and electroplated with gold. Coefficient of thermal expansion is 20.2×10^{-6}. The frequency meter design uses a cylindrical cavity operating in TE_{111} mode. The diameter of the cavity is designed for waveguide cut-off at 2 GHz. Calibration is carried out at 20°C. Calculate the error that will be introduced while measuring a frequency of 3.0 GHz at an ambient temperature of 100°C.

6.5 Consider the resonant cavity frequency meter used in Problem 6.4. Calculate the error that will be introduced at 20°C if the presence of moisture changes the value of ε_r for air to 1.002. Operating frequency is 3 GHz again.

6.6 In a transfer oscillator set-up for measurement of frequency (*see* Fig. 6.13), the variable frequency source used has a frequency range of 50 MHz to 250 MHz. The accuracy of the counter used for measuring this frequency is ± 1 Hz. What is the maximum error that could be introduced when a frequency of 10 GHz is measured?

6.7 The arrangement shown in Fig. 6.10 is used for measurement of S_{21} of a two-port network. The connector used for connecting the network to the system presents a loss which can be characterized by *S*-parameters r_{11}, r_{12}, r_{21} and r_{22} with $r_{11} = r_{22} = 0$. What will be the measured value of S_{21} in this situation?

CHAPTER 7

Passive Circuit Components

Design of passive circuit components at microwave frequencies makes use of small lengths of transmission lines or waveguides. In this chapter we shall discuss three important categories of microwave passive components: impedance transformers, filters and directional couplers.

7.1 IMPEDANCE TRANSFORMERS

An impedance transformer is perhaps the most commonly used circuit component at microwave frequencies. When two sections or components of different impedances are connected together, an impedance transformer is invariably required to ensure maximum power transfer from one part to another. A typical example is matching of a load to the characteristic impedance of the feeding line. Two techniques which are commonly employed for impedance matching make use of: (i) single or multiple stubs, and (ii) one or more quarter wave sections.

7.1.1 Use of stubs

Small sections of transmission lines connected in shunt (or series) to the main line for the purpose of impedance matching are known as stubs. Figure 7.1 shows a typical arrangement of shunt connected stub. Stubs may be open circuited or short circuited at the other end. This technique is commonly used when impedance matching over a

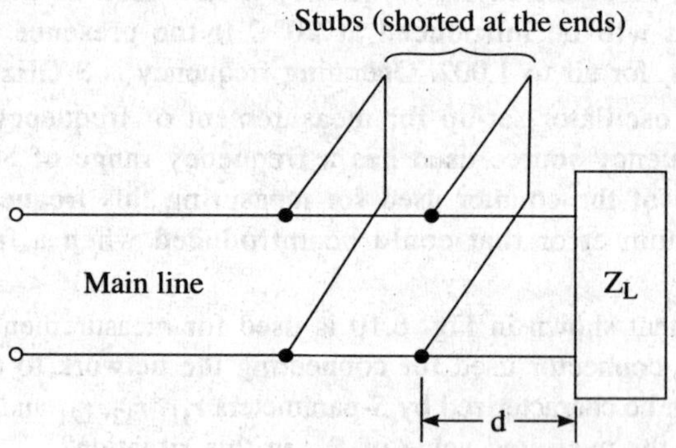

Fig. 7.1

small frequency range is required. Calculations required for design of stubs can be easily carried out by using the Smith chart.

Impedance matching can be obtined by using one or more stubs. Matching units using one, two or three stubs have been reported in literature. Calculations for the case of a single stub impedance matching transformer are shown in Fig. 7.2. Since a

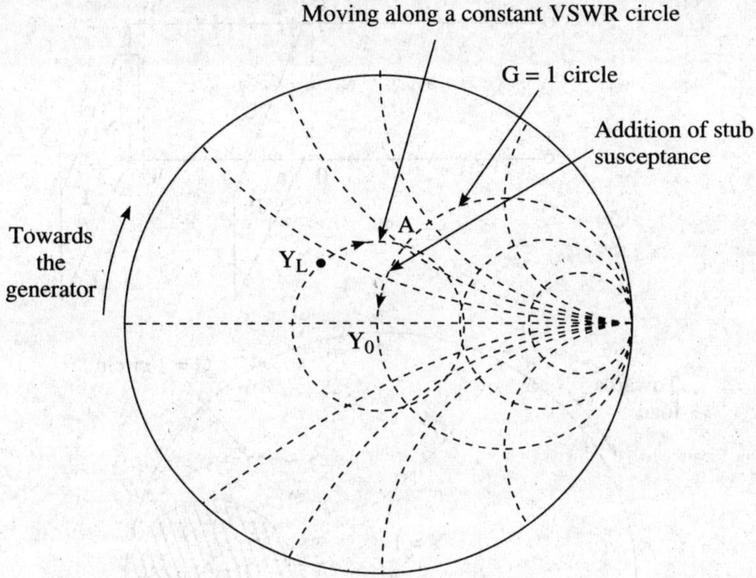

Fig. 7.2

parallel addition of reactances is involved, the Smith chart is used as an admittance chart. The distance '*d*' between the stub and the load is obtained by moving along the constant VSWR circle (towards the generator) till one reaches point *A* which lies on *G* = 1 circle. Susceptance presented by the stub should be negative of the susceptance at *A*. This determines the length of the stub.

Double stub circuit: When two stubs are used, their spacing and location along the main line can be fixed. This provides a distinct advantage in circuit fabrication and is the main reason for the popularity of this circuit. Design of double stub circuit can again be carried out by using the Smith chart.

The design procedure for the case when the distance between the two stubs is $3\lambda/8$ is illustrated in Fig. 7.3. The Smith chart is again used as an admittance chart. Stub *I* is located at the load end. For discussing this circuit, we start from the plane *oo* where the normalized admittance should be unity. Since stub *II* can add only a pure susceptance to the admittance at the junction, the admittance at the plane *aa* should lie on *G* = 1 circle. Moving a distance $3\lambda/8$ towards load, *G* = 1 circle gets transformed into another circle*. Combination of Y_L and susceptance of stub *I* should lie on this

* Shown dotted in lower half of the Smith chart in Fig. 7.3.

circle. This implies that Y_L should lie on any of the constant conductance circle intersecting this circle. Consequently, any load admittance located within the shaded area A **cannot be** matched with **this arrangement**. In order to match an admittance within **this area**, the stub I should be **located** at some distance from the load.

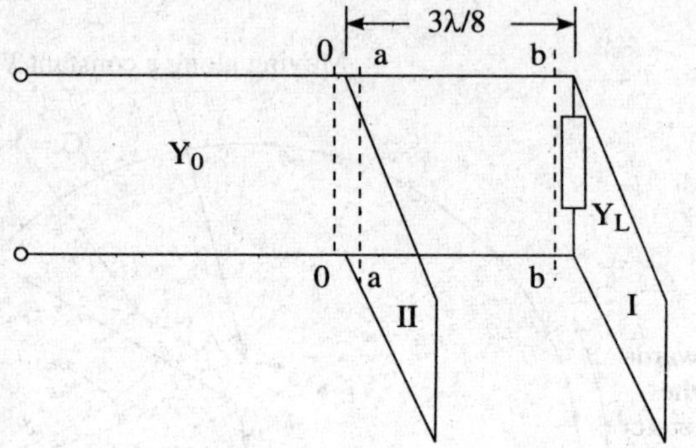

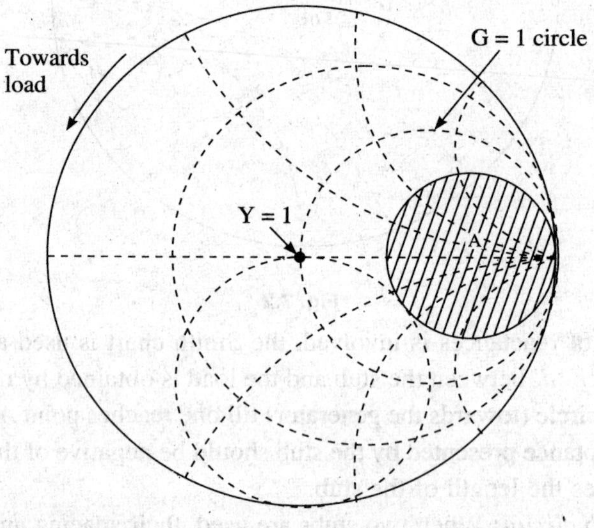

Fig. 7.3

Impedance matching using stubs suffers from the following inherent disadvantages. First, the stub circuits are essentially narrow band and their operation is limited to more or less a single frequency. Secondly, at times shunt stubs are difficult to be realized physically, particularly in waveguide circuits. These disadvantages are overcome by using multi-sectional quarter-wave transformers discussed in the following section.

7.1.2 Quarter wave transformers

A quarter wave section of a suitable characteristic impedance may be used for matching a resistive load impedance Z_L to another impedance Z_1. In terms of the arrangement shown in Fig. 7.4, equation (2.31) yields

$$Z_{in} = Z_2 \frac{Z_L + jZ_2 \tan(\beta\lambda/4)}{Z_2 + jZ_L \tan(\beta\lambda/4)} = Z_2^2 / Z_L \qquad ...(7.1)$$

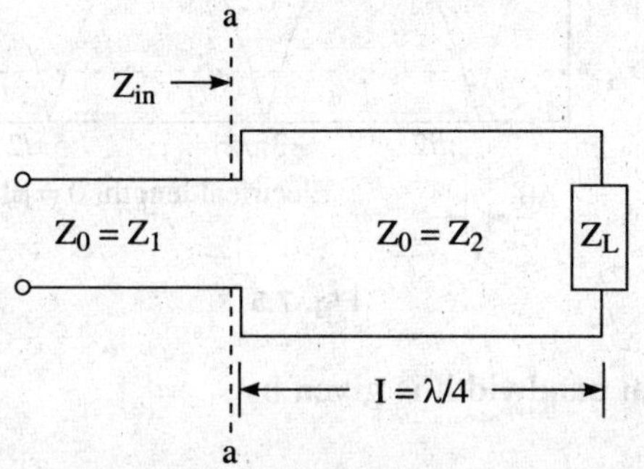

Fig. 7.4

If Z_2 is chosen equal to $\sqrt{(Z_1 Z_2)}$, the input impedance Z_{in} becomes equal to Z_1 and a perfect match is obtained at the frequency where transformer length is quarter wave length ($\lambda/4$) and at all the odd harmonics where the length corresponds to ($n\lambda/2 + \lambda/4$).

Bandwidth of this single section transformer may be evaluated by considering the reflection coefficient ρ at the input plane *aa* (Fig. 7.4). We have

$$\rho = (Z_{in} - Z_1)/(Z_{in} + Z_1) \qquad ...(7.2)$$

and $$Z_{in} = Z_2\{(Z_L + jZ_2 t)/(Z_2 + jZ_L t)\} \qquad ...(7.3)$$

where $t = \tan\beta l$. Using the condition $Z_2 = \sqrt{Z_1 Z_L}$, equation (7.2) may be written as

$$\rho = (Z_L - Z_1)/(Z_L + Z_1 + jt2\sqrt{Z_1 Z_L}) \qquad ...(7.4)$$

and $$|\rho| = 1/[1 + (2\sqrt{Z_1 Z_L}\sec\theta/(Z_L - Z_1]^2]^{1/2} \qquad ...(7.5)$$

where $\theta = \beta l$. When θ is near $\pi/2$, unity term in the denominator may be ignored and (7.5) may be approximated by

$$|\rho| = |Z_L - Z_1||\cos\theta|/(2\sqrt{Z_1 Z_L}) \qquad ...(7.6)$$

Variation of $|\rho|$ is shown in Fig. 7.5. The bandwidth or the useful frequency range of the transformer can be evaluated if the maximum permissible value of ρ within this range is specified. At ρ_m, the maximum value of reflection coefficient, the useful bandwidth provided by the transformer corresponds to the range $\Delta\theta$ shown in Fig. 7.5. The value of θ at the edge of the useful passband is given by

$$\theta_m = \cos^{-1}\left|2\rho_m\sqrt{Z_1 Z_L}/((Z_L - Z_1)\sqrt{(1-\rho_m^2)})\right| \qquad ...(7.7)$$

In the case of a TLM wave, $\theta = \beta l = \pi f/(2f_0)$, where f_0 is the frequency for which $\theta = \pi/2$. In this case the bandwidth is given by

$$\Delta f = 2(f_0 - f_m) = 2\left(f_0 - \frac{2f_0}{\pi}\theta_m\right) \qquad ...(7.8)$$

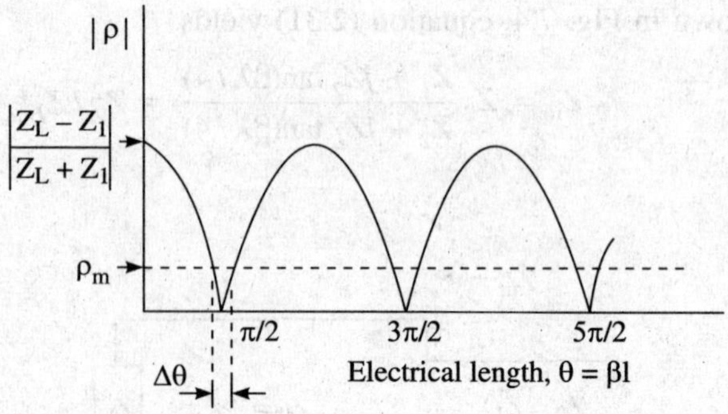

Fig. 7.5

and the fractional bandwidth is given by

$$\frac{\Delta f}{f_0} = 2-(4/\pi)\cos^{-1}\left|2\rho_m\sqrt{Z_1 Z_L}\Big/\left\{(Z_L - Z_1)\sqrt{1-\rho_m^2}\right\}\right| \qquad ...(7.9)$$

There are a number of situations where the bandwidth given by equation (7.9) is not adequate. The required increase in bandwidth, in such cases, can be obtained by using multisectional quarter wave transformers.

Before going to the analysis of multisectional transformers, two limitations of this approach should be pointed out. These are:

(i) It has been assumed that the characteristic impedance Z_1 and Z_2 are independent of frequency. For transmission lines this is a good approximation but for waveguides the impedance varies with frequency.

(ii) Changes in characteristic impedances are introduced by changes in geometrical cross-section of the transmission line or the waveguide. Fringing fields exist at these geometrical discontinuities. These discontinuity fields may be represented by a shunt susceptance at each junction. The susceptive elements modify the performance of any practical transformer from the predicted performance based on an ideal model where junction effects are neglected.

Design of multisectional impedance transformers can be simplified by assuming that individual reflection coefficients at each step are small. This small reflections approximation is discussed in the following section.

7.1.3 Small reflections approximation

When there are a number of small discontinuities along a line, the total reflection coefficient at the input may be evaluated in terms of partial reflection coefficients at various discontinuities by using small reflections approximation.

Consider the case of two discontinuities illustrated in Fig. 7.6. Partial reflection coefficients ρ_1, ρ_2 and ρ_3 are defined as

$$\rho_1 = (Z_2 - Z_1)/(Z_2 + Z_1) \qquad ...(7.10)$$

$$\rho_2 = (Z_1 - Z_2)/(Z_1 + Z_2) = -\rho_1 \qquad ...(7.11)$$

$$\rho_3 = (Z_L - Z_2)/(Z_L + Z_2) \qquad ...(7.12)$$

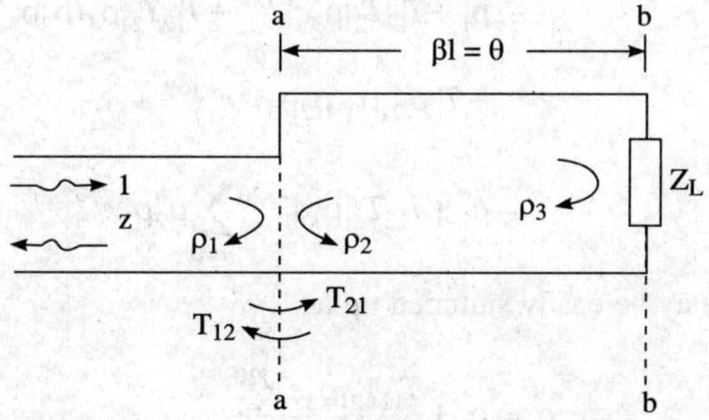

Fig. 7.6

Transmission coefficient is defined as the ratio of the amplitude of the transmitted wave to the amplitude of the incident wave and may be evaluated from the principle of conservation of power. It may be shown that for a lossless junction

$$T_{21} = 2Z_2/(Z_1 + Z_2) = 1 + \rho_1 \qquad ...(7.13)$$

$$T_{12} = 2Z_1/(Z_1 + Z_2) = 1 + \rho_2 \qquad ...(7.14)$$

Input reflection coefficient can be analysed in terms of successive reflections from two discontinuities. Various components contributing to the reflection at the input plane are shown in Fig. 7.7. When a wave of unit amplitude is incident at *aa*,

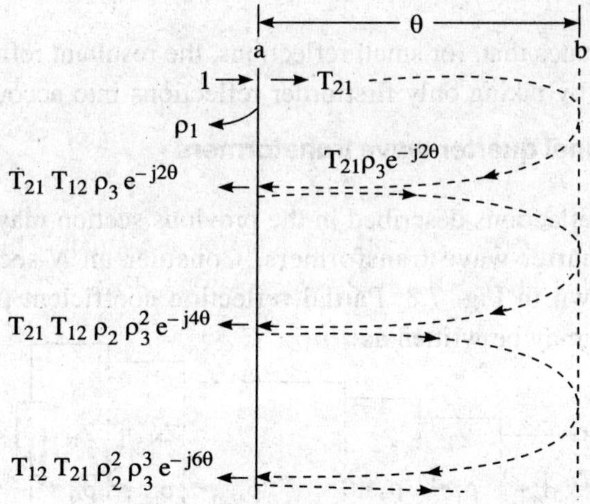

Fig. 7.7

the primary partially reflected wave has an amplitude ρ_1. As shown in Fig. 7.7, the second order reflection has an amplitude given by $T_{12}T_{21}\rho_3 e^{-j2\theta}$. Higher order successive reflections have amplitudes given by $T_{12}T_{21}\rho_3(\rho_2\rho_3)e^{-j4\theta}$, $T_{12}T_{21}\rho_3(\rho_2\rho_3)^2e^{-j6\theta}$, $T_{12}T_{21}\rho_3(\rho_2\rho_3)^4e^{-j8\theta}$etc. The total reflected wave (amplitude ρ) is given by the sum of all the partial waves sent past the plane *aa* to the left and may be written as

$$\rho = \rho_1 + T_{12}T_{21}\rho_3 e^{-j2\theta} + T_{12}T_{21}\rho_3(\rho_2\rho_3)e^{-j4\theta}$$

$$+ T_{12}T_{21}\rho_3(\rho_2\rho_3)^2 e^{-j6\theta} + \ldots$$

$$= \rho_1 + T_{12}T_{21}\rho_3 e^{-j2\theta}\sum_{n=0}^{\infty}\rho_2^n\rho_3^n e^{-j2n\theta} \qquad \ldots(7.15)$$

Series in (7.15) may be easily summed up as

$$\rho = \rho_1 + \frac{T_{12}T_{21}\rho_3 e^{-j2\theta}}{1-\rho_2\rho_3 e^{-j2\theta}} \qquad \ldots(7.16)$$

Using (7.13) and (7.14) we may write (7.16) as

$$\rho = \rho_1 + (1-\rho_1)(1-\rho_1)\rho_3 e^{-j2\theta}/(1-\rho_2\rho_3 e^{-j2\theta})$$

$$= \frac{\rho_1 + \rho_3 e^{-j2\theta}}{1+\rho_1\rho_3 e^{-j2\theta}} \qquad \ldots(7.17)$$

If ρ_1 and ρ_3 are both assumed to be small compared with unity, we may approximate (7.17) by

$$\rho = \rho_1 + \rho_3 e^{-j2\theta} \qquad \ldots(7.18)$$

This result states that, for small reflections, the resultant reflection coefficient is just that obtained by taking only first order reflections into account.

7.1.4 Multisectional quarter-wave transformers

Theory of small reflections described in the previous section may be used to analyse multisectional quarter-wave transformers. Consider an N-sectional transformer arrangement, shown in Fig. 7.8. Partial reflection coefficient ρ_n between nth and $(n + 1)$th sections may be written as

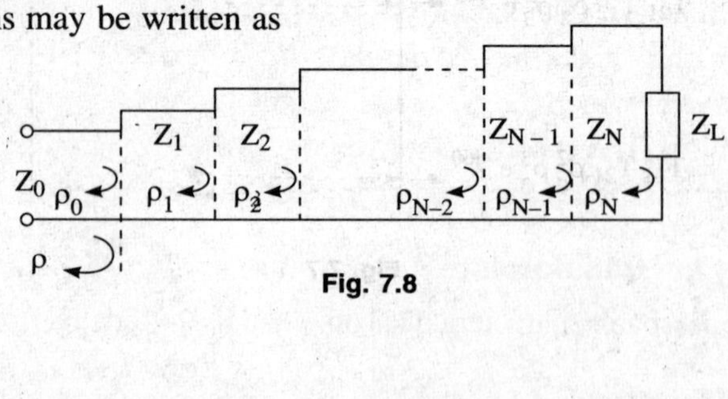

Fig. 7.8

$$\rho_n = \frac{Z_{n+1} - Z_n}{Z_{n+1} + Z_n} \quad \text{...(7.19)}$$

The last (*N*th) reflection coefficient is

$$\rho_N = \frac{Z_L - Z_N}{Z_L + Z_N}$$

where Z_L is the load impedance. All sections have equal electrical length $\beta l = \theta$ which is $\pi/2$ (quarter-wave long) at the centre frequency f_0. The load Z_L is assumed to be a pure resistance and may be greater or smaller than Z_0. In this analysis Z_L is chosen to be greater than Z_0 so that all $\rho_n = |\rho_n|$. If Z_L is smaller than Z_0, all ρ_n are negative real numbers and the only modification required to carry out the analysis is to replace ρ_n by $-\rho_n$.

Using small reflections approximation discussed in the previous section, the total reflection coefficient is obtained as the sum of first order reflected waves only. This is given by

$$\rho = \rho_0 + \rho_1 e^{-2j\theta} + \rho_2 e^{-4j\theta} \ldots + \rho_N e^{-j2N\theta} \quad \text{...(7.20)}$$

where factors $e^{-2jn\theta}$ account for the phase delays introduced because of the different distances various partially reflected waves are required to travel.

At this point it is expedient to assume that the transformer is symmetrical, that is $\rho_0 = \rho_N, \rho_1 = \rho_{N-1}, = \rho_2 = \rho_{N-2} \ldots$ etc., and therefore (7.20) becomes

$$\rho = [e^{jN\theta} + e^{-jN\theta}] + \rho_1 [e^{j(N-2)\theta} + e^{-j(N-2)\theta}] + \ldots\} \quad \text{...(7.21)}$$

where the last term is $\rho_{(N-1)/2}(e^{j\theta} + e^{-j\theta})$ when *N* is odd, and $\rho_{N/2}$ when *N* is even. It is thus seen that for a symmetrical transformer the reflection coefficient ρ is given by a Fourier Cosine series

$$\rho = 2e^{-jN\theta}\{\rho_0 \cos N\theta + \rho_1 \cos(N-2)\theta + \ldots$$
$$+ \rho_n \cos(N-2n)\theta + \ldots\} \quad \text{...(7.22)}$$

In this expression the last term is $\rho_{(N-1)/2} \cos\theta$ when *N* is odd and $\frac{1}{2}\rho N/2$ when *N* is even. It should now be apparent that by a proper choice of partial reflection coefficients ρ_n, a variety of passband characteristics can be obtained. Impedances Z_n for various selections are calculated from ρ_n's. Since the series is a cosine series, the periodic function that it defines is periodic over the interval π corresponding to frequency change over which the electrical length of each transformer section changes by a half wavelength. The specifications of coefficients ρ_n to obtain a maximally flat and an equal ripple passband characteristics are discussed in the following subsections.

7.1.5 Maximally flat transformers

A maximally flat passband characteristic is obtained if $|\rho|$ and its first $N-1$ derivatives with respect to frequency (or θ) vanish at the centre frequency f_0 (where $\theta = \pi/2$). Such a characteristic is obtained if we choose

$$\rho = A[1+e^{-j2\theta}]^N \qquad ...(7.23)$$

for which

$$|\rho| = \left|A2^N(\cos\theta)^N\right|$$

When θ = 0, we have

$$|\rho| = A2^N = \frac{Z_L - Z_0}{Z_L + Z_0}$$

and

$$A = 2^{-N}\frac{Z_L - Z_0}{Z_L + Z_0} \qquad ...(7.24)$$

Thus

$$\rho = 2^{-N}\frac{Z_L - Z_0}{Z_L + Z_0}(1+e^{-j2\theta})^N$$

$$= 2^{-N}\frac{Z_L - Z_0}{Z_L + Z_0}\sum_{n=0}^{N} C_n^N e^{-j2n\theta} \qquad ...(7.25)$$

where the binomial coefficients are given by

$$C_n^N = \frac{N(N-1)(N-2)\ldots(N-n+1)}{n!}$$

$$= \frac{N!}{(N-n)!n!} \qquad ...(7.26)$$

From small reflections approximation we have

$$\rho = \rho_0 + \rho_1 e^{-2j\theta} + \rho_2 e^{-j4\theta} + \rho_3 e^{-j6\theta} + \ldots + \rho_N e^{-j2N\theta} \qquad ...(7.27)$$

Comparing (7.25) and (7.27), and noting that $C_n^N = C_{N-n}^N$ we obtain the values of ρ_n as

$$\rho_n = 2^{-N}\frac{Z_L - Z_0}{Z_L + Z_0}C_n^N = \rho_{N-n} \qquad ...(7.28)$$

To obtain a simple solution for the characteristic impedances Z_n, it is convenient to make further approximations. Since we have already considered that all ρ_n are small, we can write

$$\ln\frac{Z_{n+1}}{Z_n} \simeq 2\frac{Z_{n+1} - Z_n}{Z_{n+1} + Z_n} = 2\rho_n \qquad ...(7.29)$$

Also we write

$$\ln\frac{Z_L}{Z_0} = 2\frac{Z_L - Z_0}{Z_L + Z_0} + \frac{2}{3}\left(\frac{Z_L - Z_0}{Z_L + Z_0}\right)^2 + \ldots \simeq 2\frac{Z_L - Z_0}{Z_L + Z_0} \qquad ...(7.30)$$

Thus we have the following relationship

$$\ln\frac{Z_{n+1}}{Z_n} = 2\rho_n = 2^{-N}C_n^N \ln\frac{Z_L}{Z_0} \qquad ...(7.31)$$

A solution for logarithm of impedances can thus be obtained. Since these values are proportional to binomial coefficients, the transformer is called a binomial transformer. As the above theory is approximate, the range of Z_L is restricted to about $0.5\ Z_0 < Z_L < 2Z_0$ for accurate results.

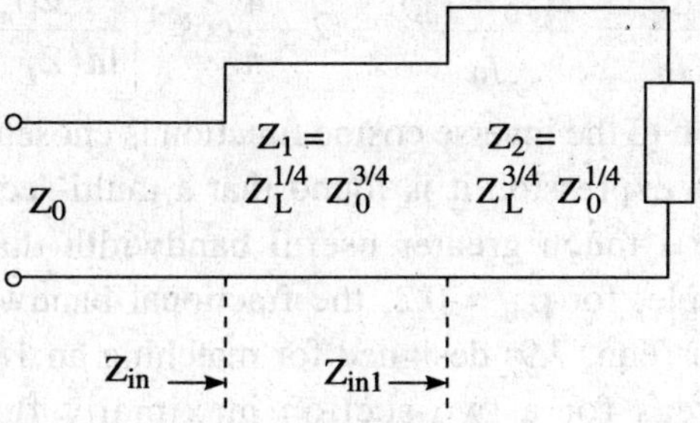

Fig. 7.9

For example, consider a two-section transformer shown in Fig. 7.9. Using (7.31) we have

$$\ln \frac{Z_1}{Z_0} = \frac{1}{4} \ln \frac{Z_L}{Z_0}$$

or
$$Z_1 = Z_L^{1/4} Z_0^{3/4} \qquad \text{...(7.32)}$$

and
$$\ln \frac{Z_2}{Z_1} = \frac{1}{2} \ln \frac{Z_L}{Z_0}$$

or
$$Z_2 = Z_L^{3/4} Z_0^{1/4} \qquad \text{...(7.33)}$$

It can be easily seen that the above values of Z_1 and Z_2 result in a perfect match at the centre frequency f_0.

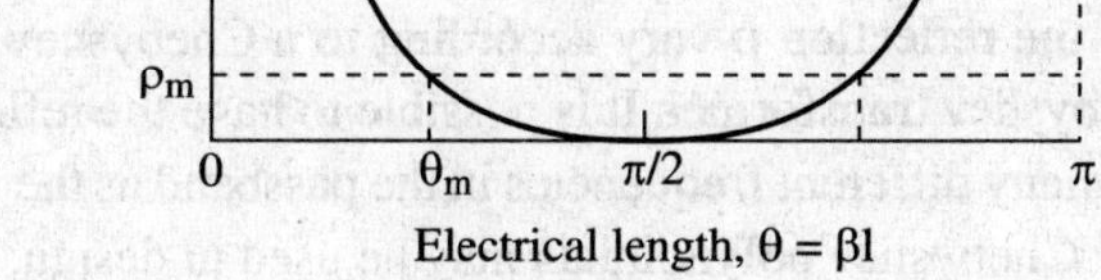

Fig. 7.10

The type of passband characteristic obtained with a maximally flat transformer is shown in Fig. 7.10. Let ρ_m be the maximum value of ρ that can be tolerated in the passband. The angle θ_m that gives $|\rho| = \rho_m$ is obtained from (7.23) and may be written as

$$\theta_m = \cos^{-1} \left| \frac{2\rho_m}{\ln(Z_L / Z_0)} \right|^{1/N} \qquad \text{...(7.34)}$$

In the case of TEM transmission line sections, $\theta = \pi f / 2f_0$, and hence the fractional bandwidth is given by

$$\frac{\Delta f}{f_0} = \frac{2(f_0 - f_m)}{f_0} = 2 - \frac{4}{\pi}\cos^{-1}\left|\frac{2\rho_m}{\ln(Z_L/Z_0)}\right|^{1/N} \qquad ...(7.35)$$

Note that the solution to the inverse cosine function is chosen such that θ_m is less than $\pi/2$. From the above expression it is found that a multi-section maximally flat transformer can provide a much greater useful bandwidth than a single section transformer. As an example, for $\rho_m = 0.2$, the fractional bandwidth obtained for a single section transformer (eqn. 7.9) designed for matching an 18 ohms load to a 50 ohms line, is 0.5 whereas for a two-section maximally flat transformer the corresponding bandwidth (eqn. 7.35) is 0.861.

7.1.6 Chebyshev transformers

Instead of a maximally flat passband characteristic, an equally useful characteristic is one that may permit the total reflection ρ to vary between 0 and ρ_m in an oscillatory manner over the passband as shown in Fig. 7.11.

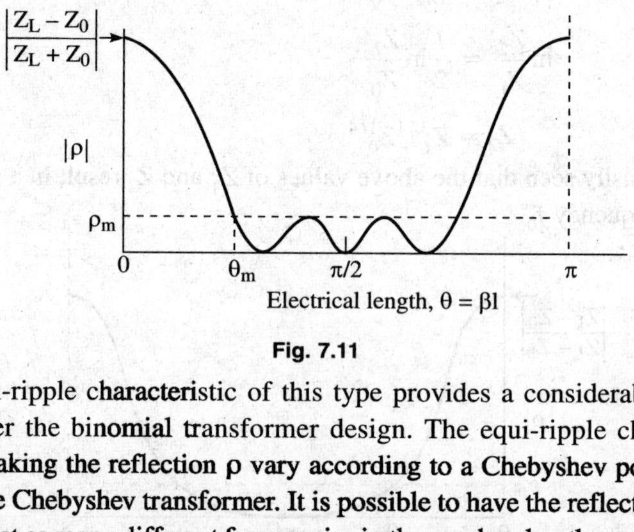

Fig. 7.11

An equal-ripple characteristic of this type provides a considerable increase in bandwidth over the binomial transformer design. The equi-ripple characteristic is obtained by making the reflection ρ vary according to a Chebyshev polynomial, and hence the name Chebyshev transformer. It is possible to have the reflection coefficient ρ become zero at as many different frequencies in the passband as there are transformer sections. To see how Chebyshev polynomials may be used in design, it is necessary to consider the basic properties of these polynomials first.

A Chebyshev polynomial of degree *n*, denoted by $T_n(x)$, is nth degree polynomial in *x*. The first four Chebyshev polynomials may be written as

$$T_1(x) = x$$

$$T_2(x) = 2x^2 - 1$$

$$T_3(x) = 4x^3 - 3x$$

$$T_4(x) = 8x^4 - 8x^2 + 1 \qquad ...(7.36)$$

A recurrence relation useful for finding nth degree polynomial from $(n - 1)$th and $(n - 2)$th degree polynomials is

$$T_n(x) = 2xT_{n-1} - T_{n-2} \qquad ...(7.37)$$

As shown in Fig. 7.12, the polynomials T_n oscillate between ± 1 for x in the range $|x| \leq 1$ and increase in magnitude monotonically for x outside this range. Chebyshev polynomials satisfy the relation

$$T_n(\cos\theta) = \cos n\theta \qquad ...(7.38)$$

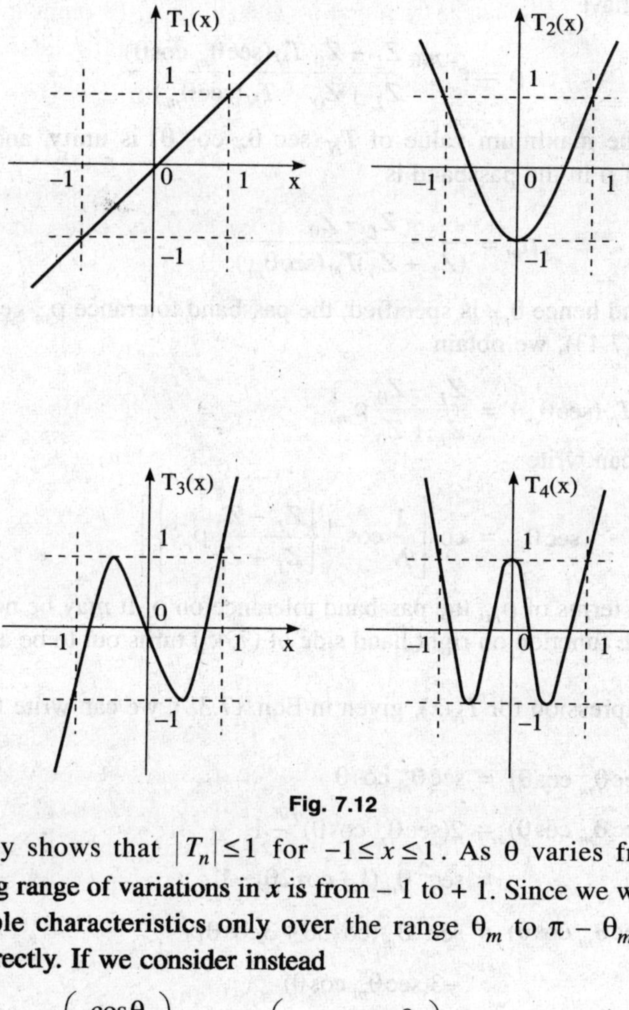

Fig. 7.12

which clearly shows that $|T_n| \leq 1$ for $-1 \leq x \leq 1$. As θ varies from 0 to π, the corresponding range of variations in x is from – 1 to + 1. Since we want $|\rho|$ to have the equi-ripple characteristics only over the range θ_m to $\pi - \theta_m$ we cannot use T_n (cos θ) directly. If we consider instead

$$T_n\left(\frac{\cos\theta}{\cos\theta_m}\right) = \cos n\left(\cos^{-1}\frac{\cos\theta}{\cos\theta_m}\right) \qquad ...(7.39)$$

we see that the argument becomes equal to unity when $\theta = \theta_m$, and less than unity for $\theta_m < \theta < \pi - \theta_m$. This function would therefore confine the equi-ripple oscillations of the polynomial T_n to the desired passband only. For a symmetrical N section transformer the reflection coefficient is given by (7.22). For a Chebyshev transformer we choose,

$$\rho = Ae^{-JN\theta}T_N(\sec\theta_m \cos\theta) \qquad ...(7.40)$$

where A is a constant. When $\theta = 0$, we have

$$\rho = \frac{Z_L - Z_0}{Z_L + Z_0} = AT_N(\sec\theta_m)$$

and therefore,

$$A = \frac{Z_L - Z_0}{(Z_L + Z_0)T_N(\sec\theta_m)} \qquad ...(7.41)$$

Consequently, we have

$$\rho = e^{-JN\theta}\frac{Z_L - Z_0}{Z_L + Z_0}\frac{T_N(\sec\theta_m \cos\theta)}{T_N(\sec\theta_m)} \qquad ...(7.42)$$

In the passband the maximum value of T_N (sec θ_m cos θ) is unity, and hence the maximum value of ρ in the passband is

$$\rho_m = \frac{Z_L - Z_0}{(Z_L + Z_0)T_N(\sec\theta_m)} \qquad ...(7.43)$$

If the passband, and hence θ_m, is specified, the passband tolerance ρ_m gets fixed and vice versa. From (7.43), we obtain

$$T_N(\sec\theta_m) = \frac{Z_L - Z_0}{Z_L + Z_0}\rho_m^{-1} \qquad ...(7.44)$$

Using (7.39), we can write

$$\sec\theta_m = \cos\left[\frac{1}{N}\cos^{-1}\left\{\frac{Z_L - Z_0}{Z_L + Z_0}\rho_m^{-1}\right\}\right] \qquad ...(7.45)$$

which given θ_m in terms of ρ_m, the passband tolerance on ρ. It may be noted that the argument of cosine function on right hand side of (7.45) turns out to be an imaginary quantity.

Using the expression for $T_N(x)$, given in Eqn. (7.38), we can write the first four polynomials as :

$$T_1(\sec\theta_m \cos\theta) = \sec\theta_m \cos\theta \qquad ...(7.46)$$

$$\begin{aligned} T_2(\sec\theta_m \cos\theta) &= 2(\sec\theta_m \cos\theta)^2 - 1 \\ &= \sec^2\theta_m(1 + \cos 2\theta) - 1 \end{aligned} \qquad ...(7.47)$$

$$\begin{aligned} T_3(\sec\theta_m \cos\theta) &= \sec^3\theta_m(\cos 3\theta + 3\cos\theta) \\ &\quad -3(\sec\theta_m \cos\theta) \end{aligned} \qquad ...(7.48)$$

$$\begin{aligned} T_4(\sec\theta_m \cos\theta) &= \sec^4\theta_m(\cos 4\theta + 4\cos 2\theta + 3) \\ &\quad -4\sec^2\theta_m(\cos 2\theta + 1) \end{aligned} \qquad ...(7.49)$$

These results are sufficient for designing transformers up to four sections in length. A greater number of sections would rarely be required in practice. Values of ρ_n **are obtained by comparing (7.22) with (7.40) with** T_n **given by** (7.47), **(7.48) and** (7.49) for two, three and four section transformer, respectively. As an example, consider the design of a two-section transformer to match a line with $Z_0 = 1$ to a line or load with $Z_L = 2$. Let the maximum tolerable value of ρ be $\rho_m = 0.05$. Using Eqn. (7.43) we obtain,

$$T_2(\sec\theta_m) = 2\sec^2\theta_m - 1 = \frac{1}{3(0.05)} = 6.67$$

and hence sec $\theta_m = 1.96$ and $\theta_m = 1.04$. Thus the fractional bandwidth is obtained as

$$\frac{\Delta\theta}{\pi/2} = \frac{\Delta f}{f_0} = \frac{4}{\pi}\left(\frac{\pi}{2} - 1.04\right) = 0.675.$$

From (7.22) and (7.47)

$$\begin{aligned} 2\rho_0 \cos 2\theta + \rho_1 &= \rho_m T_2(\sec\theta_m \cos\theta) \\ &= \rho_m \sec^2\theta_m \cos 2\theta + \rho_m(\sec^2\theta_m - 1) \end{aligned} \quad ...(7.50)$$

and hence

$$\rho_0 = \frac{1}{2}\rho_m \sec^2\theta_m = \rho_2 = 0.096$$

$$\rho_1 = \rho_m(\sec^2\theta_m - 1) = 0.142 \quad ...(7.50a)$$

Impedances Z_1 and Z_2 are given by

$$Z_1 = \frac{1+\rho_0}{1-\rho_0} Z_0 = 1.21$$

$$Z_2 = \frac{1+\rho_1}{1-\rho_1} Z_1 = 1.62$$

A plot of the passband characteristic in this case is shown is Fig. 7.13.

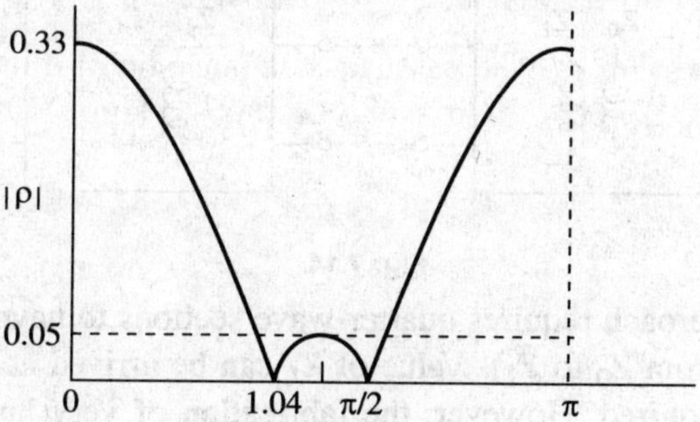

Fig. 7.13

7.2 MICROWAVE FILTERS

Methods for designing filters at low frequencies are partly applicable to the design of filters at microwave frequencies. Low frequency filter design consists of the following steps : (i) Design of a prototype low-pass filter with the desired maximally flat or equi-ripple characteristics, (ii) Transformation of this prototype network to the required type (low-pass, high-pass, band-pass or band-elimination) filter with the specified centre and/or edge frequencies; and (iii) realization of the network in terms of circuit elements (inductors and capacitors). The first two steps are identical for filter design at microwave frequencies also. Realization of the network is different because, mostly, the microwave circuits use distributed elements and not the lumped inductors and capacitors. However, during past few years lumped inductors and capacitors have been fabricated and used at microwave frequencies also. Use of lumped elements at microwave frequencies will be discussed later in Chapter 11. If lumped elements are used for filter circuits at microwave frequencies, microwave filter design becomes totally identical to the low frequency filter design.

There is an alternative approach for designing filters at microwave frequencies. Multisectional quarter wave transformers designed for impedance-matching possess band-pass characteristics. In fact these transformers can be used as band-pass filters if the different impedance levels at the two ends are acceptable for the circuits that precede and follow the filter stage. Even where equal impedance levels are desired at the input and the output of the filter, the transformer design could be modified to design filters.

The following two sub-sections describe conversion of impedance matching transformers to filters and realization of microwave filters from low frequency prototype circuits.

7.2.1 Filters derived from multisectional quarter wave transformers

If two multisectional impedance transformers (Z_0 to Z_L) are connected back to back as shown in Fig. 7.14, the composite network will have input and output impedances

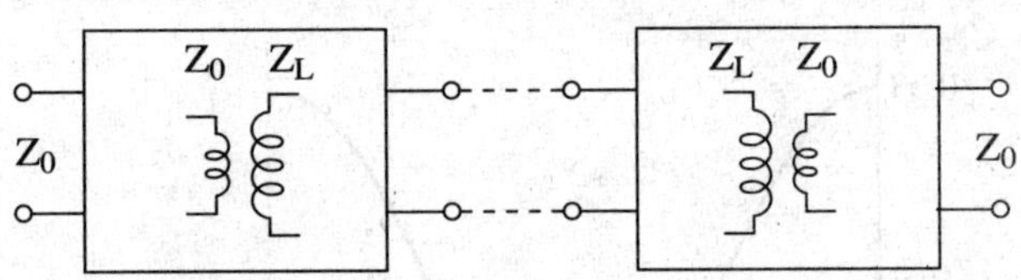

Fig. 7.14

equal to Z_0. This approach requires quarter-wave sections to have impedance values over a wide range (from Z_0 to Z_L). Value of Z_L can be arrived at by considering the off-band rejection required. However, the fabrication of very large and very small impedance sections becomes difficult because of the mechanical tolerances required. Also, there are discontinuity reactances that should be considered wherever the ratio of the impedances of the adjacent sections becomes large.

There is an alternative approach wherein the impedances are stepped up and stepped down alternately. This method is based on the equivalence between an impedance step-up and an equal impedance step-down with a quarter wave a transformer preceding the step-down. This is illustrated by two circuits shown in Figs. 7.15 (a) and (b). We note that in both the cases, the reflection coefficient is

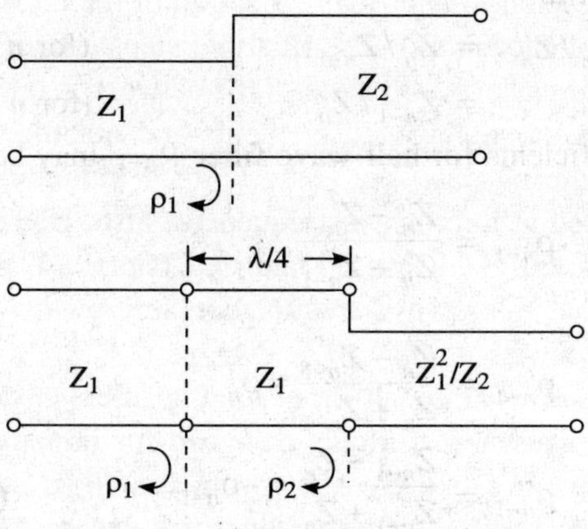

Fig. 7.15

given by $(Z_2 - Z_1)/(Z_2 + Z_1)$. Using this arrangement a multisectional quarter wave transformer can be converted into a multisectional half wave filter. This transformation is shown in Fig. 7.16. Impedance of the quarter wave transformer section, Z_1, Z_2, Z_3

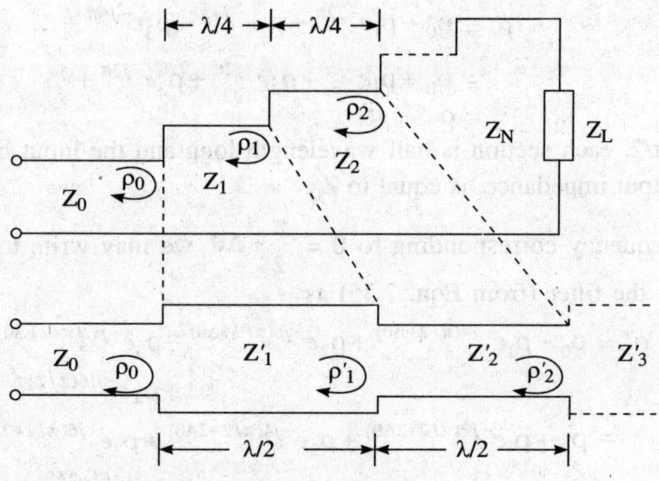

Fig. 7.16

etc., are related to the impedances Z'_1, Z'_2, Z'_3, etc., of half-wave filter sections as follows:

$$Z_1' = Z_1$$
$$Z_2' / Z_1' = Z_1 / Z_2$$
$$Z_3' / Z_2' = Z_3 / Z_2$$
$$Z_4' / Z_3' = Z_3 / Z_4$$
$$Z_5' / Z_4' = Z_5 / Z_4$$

In general, we may write

$$Z'_n / Z'_{n-1} = Z_n / Z_{n-1} \ldots \quad \text{(for } n \text{ odd)}$$
$$= Z_{n-1} / Z_n \ldots \quad \text{(for } n \text{ even)} \qquad \ldots(7.51)$$

Partial reflection coefficients for half wave filter ρ'_{n-1} may be written as

$$\rho'_{n-1} = \frac{Z'_n - Z'_{n-1}}{Z'_n + Z'_{n-1}} \qquad \ldots(7.52)$$

Using (7.51) we have

$$\rho'_{n-1} = \frac{Z_n - Z_{n-1}}{Z_n + Z_{n-1}} = \rho_{n-1} \quad \text{(for } n \text{ odd)} \qquad \ldots(7.53)$$

and

$$\rho'_{n-1} = \frac{Z_{n-1} - Z_n}{Z_{n-1} + Z_n} = -\rho_{n-1} \quad \text{(for } n \text{ even)} \qquad \ldots(7.54)$$

Total reflection coefficient for the half wave filter may be written as

$$\rho' = \rho'_0 + \rho'_1 e^{-j4\theta} + \rho'_2 e^{-j8\theta} + \rho'_3 e^{-j12\theta} + \ldots \qquad \ldots(7.55)$$

where 2θ is the electrical length (in radians) of each half wave filter section. At $\theta = \pi/2$, that is at mid-band,

$$\rho' = \rho_0 - \rho_1 e^{-j2\pi} + \rho_2 e^{-j4\pi} - \rho_3 e^{-j6\pi} + \ldots$$
$$= \rho_0 + \rho_1 e^{-j\pi} + \rho_2 e^{-j2\pi} + \rho_3 e^{-j3\pi} + \ldots$$
$$= \rho \qquad \ldots(7.56)$$

At $\theta = \pi/2$, each section is half wavelength long and the input impedance, as well as the output impedance, is equal to Z_0.

For a frequency corresponding to $\theta = \frac{\pi}{2} + \Delta\theta$ we may write total reflection coefficient for the filter (from Eqn. 7.55) as

$$\rho' = \rho_0 - \rho_1 e^{-j4(\pi/2+\Delta\theta)} + \rho_2 e^{-j8(\pi/2+\Delta\theta)} - \rho_3 e^{-j12(\pi/2+\Delta\theta)} + \rho_4 e^{-j16(\pi/2+\Delta\theta)} + \ldots$$
$$= \rho_0 + \rho_1 e^{-j2(\pi/2+2\Delta\theta)} + \rho_2 e^{-j4(\pi/2+2\Delta\theta)} + \rho_3 e^{-j6(\pi/2+2\Delta\theta)} + \rho_4 e^{-j8(\pi/2+2\Delta\theta)} + \ldots \qquad \ldots(7.57)$$

Substituting $\theta' = \frac{\pi}{2} + 2\Delta\theta$, we have

$$\rho' = \rho_0 + \rho_1 e^{-j2\theta'} + \rho_2 e^{-j4\theta'} + \rho_3 e^{-j6\theta'} + \ldots \qquad \ldots(7.58)$$

Equation (7.58) indicates that the frequency response of the filter with respect to parameter θ' is identical to that of the original transformer with respect to parameter θ. Since $(\theta' - \pi/2)$ is twice of $(\theta - \pi/2)$ the passband of the filter should be half that of the transformer.

Thus we conclude that by using the method outlined above a multi-sectional quarter wave transformer may be transformed into a half wave filter with half the bandwidth.

7.2.2 Filters based on low frequency prototype design

As mentioned earlier microwave filters may also be derived from low frequency prototype filter design. These filter design methods are available in several books on electrical networks and will not be discussed here. Low frequency filters are designed in the form of ladder structures of lumped reactances shown in Fig. 7.17 for various types of filters. Basic components of these networks are: (a) inductors, (b) capacitors,

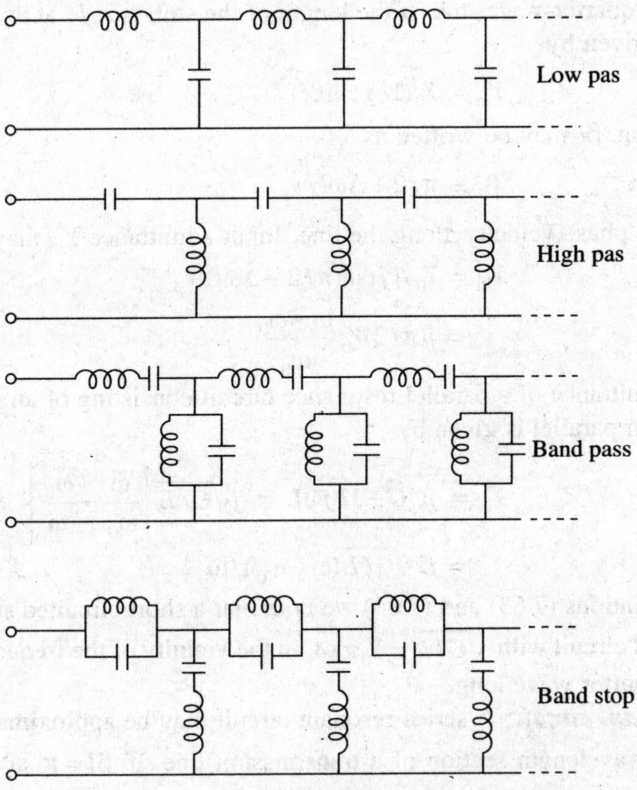

Fig. 7.17

(c) parallel resonant circuits and (d) series resonant circuits. At microwave frequencies these elements may be realized by making use of transmission line sections as discussed below:

a. ***Inductors:*** An inductor may be realized by a section of transmission line (length l less than quarter wavelength) shorted at one end. Input impedance of such a line is

$$Z_{in} = jZ_0 \tan\beta l$$

and equivalent inductance may be written as

$$L = \frac{Z_0}{\omega}\tan\beta l \qquad ...(7.59)$$

b. Capacitors: A capacitor may also be realized by a shorted transmission line provided its length is greater than quarter wavelength and less than half the wavelength. In this case,

$$Z_{in} = jZ_0 \tan\beta l, \qquad (\pi/2 < \beta l < \pi)$$
$$= -jZ_0 \cot(\beta l - \pi/2)$$

and
$$C = \frac{1}{\omega Z_0}\tan(\beta l - \pi/2) \qquad ...(7.60)$$

c. Parallel resonant circuit: A parallel resonant circuit may be approximated by a short circuited quarter-wave stub. If the length of the stub is $\lambda_0/4$ at $\omega = \omega_0$, the input admittance is given by

$$Y_{in} = Y_0(1/j\tan\beta l) \qquad ...(7.61)$$

At $\omega = \omega_0 + \Delta\omega$, βl may be written as

$$\beta l = \pi/2 + \Delta\omega l/v_p \qquad ...(7.62)$$

where v_p is the phase velocity along the line. Input admittance Y_{in} may be written as

$$Y_{in} = Y_0/[j\tan(\pi/2 + \Delta\omega l/v_p)]$$
$$\simeq j(\pi/2)Y_0\frac{\omega - \omega_0}{\omega_0} \qquad ...(7.63)$$

Input admittance of a parallel resonance circuit consisting of an inductor L and a capacitor C in parallel is given by

$$Y_{in} = j\omega C + 1/j\omega L = j\sqrt{C/L}\left\{\frac{\omega}{\omega_0} - \frac{\omega_0}{\omega}\right\}$$
$$\simeq j2\sqrt{C/L}(\omega - \omega_0)/\omega_0 \qquad ...(7.64)$$

Comparing equations (7.63) and (7.64) we note that a short circuited stub behaves as a parallel-tuned circuit with $\sqrt{C/L} = Y_0\pi/4$ in the vicinity of the frequency for which the stub is a quarter wave long.

d. Series resonant circuit: A series resonant circuit may be approximated by a short circuited half wavelength section of a transmission line. If $\beta l = \pi$ at $\omega = \omega_0$, input impedance at $\omega = \omega_0 + \Delta\omega$ may be written as

$$Z_{in} = jZ_0(\pi/2)(\omega - \omega_0)/\omega_0 \qquad ...(7.65)$$

whereas, the input impedance of a series resonant circuit near resonance is given by

$$Z_{in} = j2\sqrt{L/C}(\omega - \omega_0)/\omega_0 \qquad ...(7.66)$$

Comparing equations (7.65) and (7.66) we note that a short circuited half wavelength stub behaves as a series tuned circuit with $\sqrt{L/C} = Z_0(\pi/4)$ in the vicinity of frequency for which it is a half wave long.

Thus various elements of filters shown in Fig. 7.17 may be realized in terms of shorted transmission line stubs of various lengths. However, in addition to realization of various elements, the ladder structure of Fig. 7.17 is itself not very convenient for realization in terms of transmission line or waveguide sections. This ladder structure may be converted into a periodically loaded transmission line network as discussed below.

Consider a section of ladder network and a $\lambda/2$ long section of loaded transmission line as shown in Fig. 7.18. Characteristic impedance of the line is taken as unity. It can be seen that these two networks have identical impedance transformation properties. For example, if the two networks are terminated in an impedance Z_x at the output, the input impedance for both the cases is $Z_2 + Z_1 Z_x/(Z_1 + Z_x)$. Similarly, if the two networks are terminated with an impedance Z_y at the input, the impedance at output terminals is $Z_1(Z_2 + Z_y)/(Z_1 + Z_2 + Z_y)$. Thus the two structures are identical and a transmission line with shunt stubs spaced quarter-wavelength apart may be used for realizing the filters shown in Fig. 7.17 at microwave frequencies.

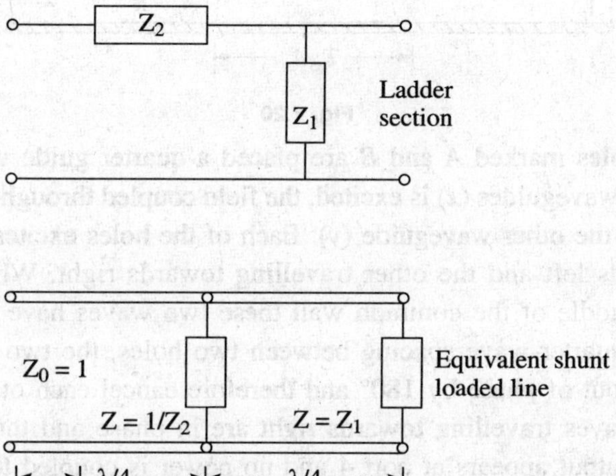

Fig. 7.18

7.3 DIRECTIONAL COUPLERS

A directional coupler is a 4-port network commonly used for sampling a known fraction of the microwave power flowing in a particular direction. Figure 7.19 represents a directional coupler if, when a wave travels from port 1 to port 2, a fixed fraction of this power appears at port 4 and there is no output at port 3. Conversely, if the wave is travelling from port 2 to port 1, a fraction of this signal appears at port 3 and there is no output at port 4. When the transmission takes place from port 1 to port 2, the port 4 is called coupled port and the port 3 is called *isolated port.*

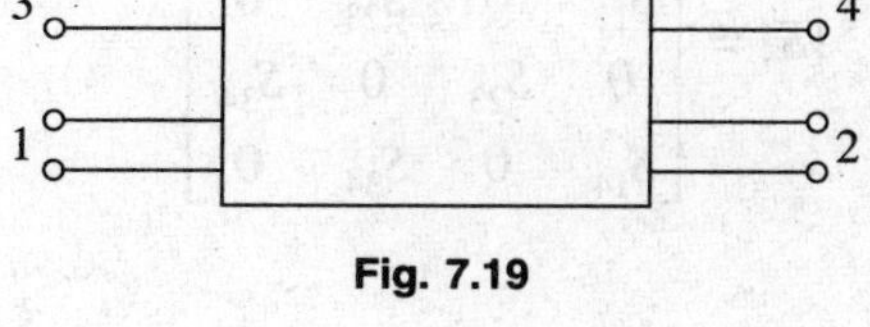

Fig. 7.19

7.3.1 Multihole directional coupler

The most common type of the waveguide directional coupler consists of two waveguides placed adjacent to each other such that they share a common wall. There are a number of holes in the common wall to provide coupling from one guide to the another. The principle of such a multihole directional coupler can be illustrated by considering two holes as shown in Fig. 7.20.

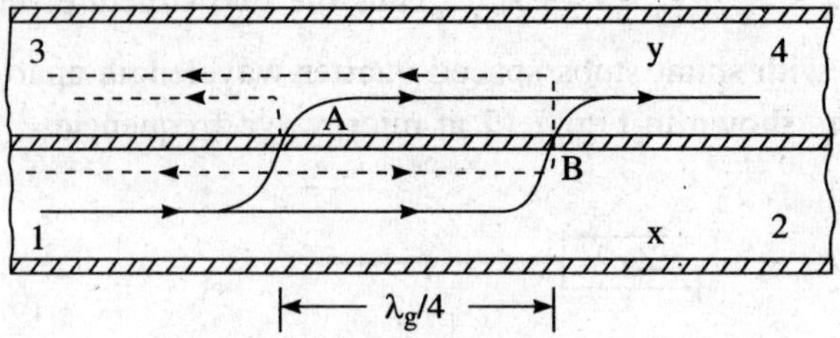

Fig. 7.20

The two holes marked *A* and *B* are placed a quarter guide wavelength apart. When one of the waveguides (*x*) is excited, the field coupled through each of the holes excites waves in the other waveguide (*y*). Each of the holes excites two waves, one travelling towards left and the other travelling towards right. When the holes are located in the middle of the common wall these two waves have equal amplitude. Because of the quarter-wave spacing between two holes, the two waves travelling towards left are out of phase by 180° and therefore cancel each other. On the other hand, the two waves travelling towards right are in phase and thus reinforce each other. Thus an output appears at port 4 and no power is coupled to the port 3. The amount of power coupled to the port 4 depends upon the sizes and the locations of the holes in the common wall.

The two-hole coupler described above works satisfactorily only at a single frequency corresponding to quarter-wave spacing between the holes. As in the case of transformer discussed in Sec. 7.1, the bandwidth of a directional coupler can also be increased by using multiple pairs of coupling holes. In this case also, both maximally flat aud equi-ripple responses can be designed for.

7.3.2 Scattering matrix for directional couplers

When all the four ports of a directional coupler are matched, the scattering matrix for the network may be written as

$$[S] = \begin{bmatrix} 0 & S_{12} & 0 & S_{14} \\ S_{12} & 0 & S_{23} & 0 \\ 0 & S_{23} & 0 & S_{34} \\ S_{14} & 0 & S_{34} & 0 \end{bmatrix} \qquad ...(7.67)$$

While writing eqn. (7.67), complete isolation between ports 1 and 3 and between ports 2 and 4 has been considered. Also the reciprocal nature of the network has been taken into account. Equation (7.67) may be simplified by considering the properties of scattering matrices for lossless networks. Using (5.43) we have

$$S_{12}S_{12}^* + S_{14}S_{14}^* = 1 \qquad ...(7.68)$$

$$S_{12}S_{12}^* + S_{23}S_{23}^* = 1 \qquad ...(7.69)$$

$$S_{23}S_{23}^* + S_{34}S_{34}^* = 1 \qquad ...(7.70)$$

$$S_{14}S_{14}^* + S_{34}S_{34}^* = 1 \qquad ...(7.71)$$

From (7.68) and (7.69) we get

$$|S_{14}| = |S_{23}| \qquad ...(7.72)$$

and from (7.69) and (7.70) we may write

$$|S_{12}| = |S_{34}| \qquad ...(7.73)$$

At this stage we make use of an artifice which is commonly employed to change the phase angles of the coefficients of a scattering matrix. As discussed in Sec. 5.5.1 (and also in Problem 5.4), phase angle of S_{mm}'s may be changed by adding sections of line (or waveguide) at various ports. Alternatively, reference planes for various ports may be redefined. By choosing reference plane of port 1 with respect to the reference plane of port 2 we can make S_{12} real. In the same manner S_{31} can be made real by selecting suitably the reference plane of port 3 with reference to that of port 4. Thus,

$$S_{12} = S_{34} = \alpha \qquad ...(7.74)$$

Using the second characteristics of lossless networks (5.48) we can write

$$S_{12}S_{23}^* + S_{14}S_{12}^* = 0 \qquad ...(7.75)$$

Since S_{12} is real

$$S_{23}^* + S_{14} = 0 \qquad ...(7.76)$$

We can further select reference plane of port 4 with respect to port 1 such that S_{14} is real. In that case one can write

$$S_{23} = -S_{14} = \beta \qquad ...(7.77)$$

The scattering matrix (7.67) may now be written as

$$[S] = \begin{bmatrix} 0 & \alpha & 0 & -\beta \\ \alpha & 0 & \beta & 0 \\ 0 & \beta & 0 & \alpha \\ -\beta & 0 & \alpha & 0 \end{bmatrix} \qquad ...(7.78)$$

Coefficients α and β are such that

$$\alpha^2 + \beta^2 = 1 \qquad ...(7.79)$$

α is called the transmission factor and β is known as the *coupling factor.* For a 20 dB directional coupler we have

$$\beta^2 = -20 \text{ dB} = 0.01, \quad \text{i.e. } \beta = 0.1 \text{ and } \alpha = \sqrt{99/100} \simeq 0.995.$$

7.3.3 Coupled line directional couplers

In addition to the multihole directional couplers discussed earlier, coupled line directional couplers are also used extensively. These couplers make use of continuous coupling between two lines placed close to each other. This technique is very well suited for stripline and microstrip transmission lines. These transmission lines will be described in detail in Chapter 10 when we discuss microwave integrated circuits. However, the

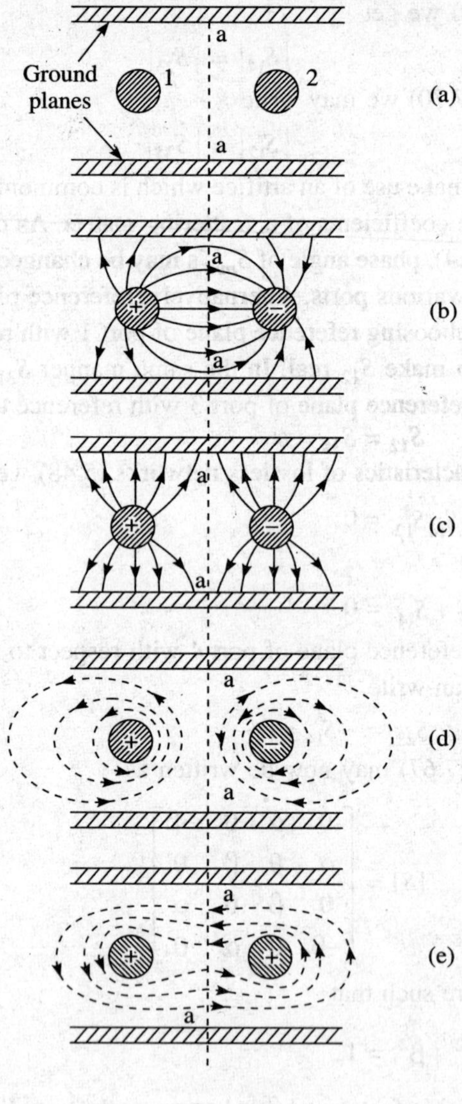

Fig. 7.21

concept of coupled lines can be introduced by considering the configuration shown in Fig. 7.21. Basic transmission line configuration shown, i.e. central conductor surrounded by two planar ground slabs is called a slab line. When we have two conductors as shown in Fig. 7.21 (a), it forms a coupled line system. It may be noted that the two lines marked 1 and 2 are identical and a plane of geometrical symmetry exists at *aa.* In general, a coupled line system (similar to that shown in Fig. 7.21 a) can support two modes. The electric field configurations for the two modes are shown in Figs. 7.21 (b) and (c), respectively. For the mode shown in Fig. 7.21 (b), a majority of the electric field lines start at the conductor 1 and terminate at the conductor 2. The instantaneous voltages at two conductors are out of phase by 180°. All electric field lines cross the plane of symmetry *aa* at right angles and there is no component of the electric field tangential to the plane of symmetry. Thus *aa* can be replaced by an electric wall. This mode is known as ***odd mode.*** Magnetic field configuration for this mode is shown in Fig. 7.21 (d). Magnetic field lines are completely tangential to the plane *aa* and do not cross this plane. The electric field configuration for the other mode called the *even mode,* is shown in Fig. 7.21 (c). In this case both the conductors 1 and 2 are in phase and the field lines originate from these two conductors and terminate on the two ground slabs. There are no *E*-field lines going from conductor 1 to 2 (or vice versa). Again the field is symmetrical about the plane *aa* but now this plane acts as a magnetic wall with electric field lines being completely tangential to this plane. The magnetic field configuration in this case is shown in Fig. 7.21 (e). In general, the characteristic impedances for the two modes of the coupled line system are different. These are denoted by Z_{00} and Z_{0e} for the odd mode and the even mode, respectively.

Even and odd modes of wave propagation along coupled lines are significant in the sense that any arbitrary excitation of the coupled line system may be described in terms of a suitable linear combination of these even and odd modes. This representation is shown in Fig. 7.22 for the case when two arbitrary voltages V_1 and V_2 are applied to ports 1 and 2 of the four-port network which possesses a plane of symmetry as in the case of coupled line system. Response for the excitation in Fig. 7.22(a) can be obtained by superimposing the responses obtained for even mode excitation (Fig. 7.22b) and odd mode excitation (Fig. 7.22c). For even or odd mode excitations (shown in Fig. 7.22b and 7.22c, respectively) symmetry considerations allow the four-port network to be treated as a two-port network as shown in these figures. Let reflection and transmission coefficients for the odd mode be ρ_{00} and T_{00}, and those for the even mode be ρ_{0e} and T_{0e} respectively. Reflected signals from the four ports when $V_1 = 1$ and $V_2 = 0$ (Fig. 7.22d) may be written as

$$b_1 = \frac{1}{2}(\rho_{00} + \rho_{0e}) \quad ...(7.80)$$

$$b_2 = \frac{1}{2}(\rho_{00} - \rho_{0e}) \quad ...(7.81)$$

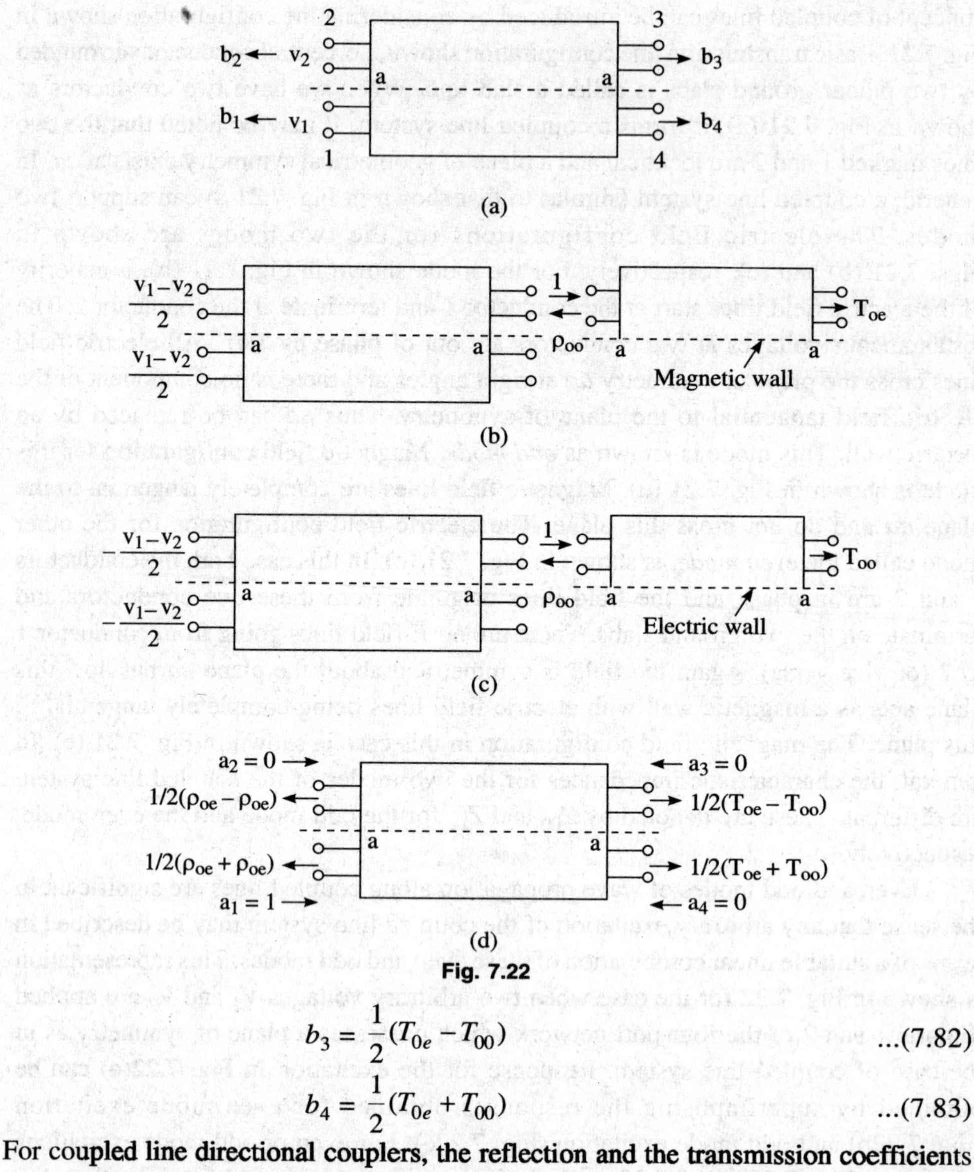

Fig. 7.22

$$b_3 = \frac{1}{2}(T_{0e} - T_{00}) \qquad \text{...(7.82)}$$

$$b_4 = \frac{1}{2}(T_{0e} + T_{00}) \qquad \text{...(7.83)}$$

For coupled line directional couplers, the reflection and the transmission coefficients for the even and the odd modes may be evaluated in terms of characteristic impedances and electrical lengths of the coupled lines for the two modes. For a transmission line section of characteristic impedance Z **and electrical length** θ **(Fig. 7.23), the reflection and the transmission coefficients are given by**

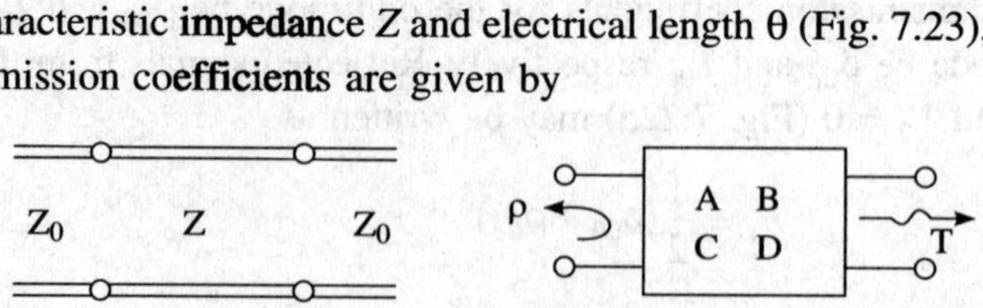

Fig. 7.23

$$\rho = \frac{A + B/Z_0 - CZ_0 - D}{A + B/Z_0 + CZ_0 + D} \quad ...(7.84)$$

$$T = 2/(A + B/Z_0 + CZ_0 + D) \quad ...(7.85)$$

The coefficients *A, B, C* and *D* are given by

$$\begin{bmatrix} A & B \\ C & D \end{bmatrix} = \begin{bmatrix} \cos\theta, & jZ\sin\theta \\ (j/Z)\sin\theta, & \cos\theta \end{bmatrix} \quad ...(7.86)$$

where Z is the characteristic impedance of the transmission line section and Z_0 is the impedance of lines connected at the two ports.

Substituting (7.86) in (7.84) and (7.85) we have

$$\rho = \frac{j[(Z/Z_0) - (Z_0/Z)]\sin\theta}{\Sigma} \quad ...(7.87)$$

$$T = 2/\Sigma \quad ...(7.88)$$

where $$\Sigma = 2\cos\theta + j[(Z/Z_0) + (Z_0/Z)]\sin\theta \quad ...(7.89)$$

Equations (7.87) and (7.88) may be used to write the reflection and the transmission coefficients for the even and the odd modes by replacing Z with the characteristic impedance for these modes and θ with electrical lengths for these modes. Interesting results are exhibited when

$$\theta = \theta_e = \theta_0 \quad ...(7.90)$$

and $$Z_0 = (Z_{0e}Z_{00})^{1/2} \quad ...(7.91)$$

Relation (7.90) implies that phase velocities for the even and the odd modes are equal. This is true for the coupled line configuration shown in Fig. 7.21 and for some other lines (striplines, for example). However, there are certain transmission structures (such as microstrips discussed in Chapter 10) where this condition is not strictly valid. Equation (7.91) defines the characteristic impedance of the lines connected at the four output ports of a coupled line system. When (7.90) and (7.91) hold, we have

$$\rho_{0e} = -\rho_{00} = j[(Z_{0e}/Z_{00})^{1/2} - (Z_{00}/Z_{0e})^{1/2}](\sin\theta)/\Sigma \quad ...(7.92)$$

$$T_{0e} = T_{00} = 2/\Sigma \quad ...(7.93)$$

where $$\Sigma = 2\cos\theta + j[(Z_{0e}/Z_{00})^{1/2} + (Z_{00}/Z_{0e})^{1/2}]\sin\theta \quad ...(7.94)$$

Using these results to evaluate b's for Fig. 7.22(d), we have

$$b_1 = 0 \quad ...(7.95)$$

$$b_2 = \rho_{0e} = j[Z_{0e}/Z_{00})^{1/2} - (Z_{00}/Z_{0e})^{1/2}](\sin\theta)/\Sigma \quad ...(7.96)$$

$$b_3 = 0 \quad ...(7.97)$$

$$b_4 = T_{0e} = 2/\Sigma \quad ...(7.98)$$

When $\theta = \pi/2$, i.e. coupled lines are quarter wavelength long, the expressions for b_2 and b_4 may be simplified as :

$$b_2 = [(Z_{0e}/Z_{00})^{1/2} - (Z_{00}/Z_{0e})^{1/2}]/[(Z_{0e}/Z_{00})^{1/2} + (Z_{00}/Z_{0e})^{1/2}]$$

$$= [(Z_{0e}/Z_{00}) - 1]/[(Z_{0e}/Z_{00}) + 1] \qquad ...(7.99)$$

$$= \beta$$

$$b_4 = -2/[j\{(Z_{0e}/Z_{00})^{1/2} + (Z_{00}/Z_{0e})^{1/2}\}] \qquad ...(7.100)$$

Using (7.99) b_4 may be written as

$$b_4 = -j\sqrt{1-\beta^2},$$

and

$$|b_4| = \sqrt{1-\beta^2} = \alpha \qquad ...(7.101)$$

Equation (7.95) shows that the input port is matched perfectly whereas (7.97) shows that port 3 is completely isolated. Port 2 is the coupled port and β (as in 7.77 and 7.78) is the coupling factor. Port 4 is directly connected to port 1 and α is the transmission factor. Outputs at various ports when a signal of unit amplitude is fed to the port 1 are shown in Fig. 7.22 (d).

Thus we conclude that a section of coupled pair of lines acts as a perfect directional coupler provided equations (7.90) and (7.91) are satisfied. Under these conditions coupling to ports 2 and 4 is given by

$$b_2 = \rho_{0e} = \beta \qquad ...(7.102)$$

and

$$b_4 = T_{0e} = -j\sqrt{1-\beta^2}$$

MULTI-SECTIONAL COUPLERS

As in the case of quarter-wave transformers the bandwidth of coupled line directional couplers can also be increased by cascading multiple sections of coupled lines. For analysing the multiple section couplers, the four-port directional coupler is compared to a two-port network shown in Fig. 7.23. It can be verified that when a transmission line of impedance Z_{0e} and length $\lambda/4$ is connected in between transmission lines of impedance Z_0, reflection and transmission coefficients are equal to the outputs at port 2 and port 4, respectively. This representation (Fig. 7.24) simplifies the analysis of multisectional directional couplers.

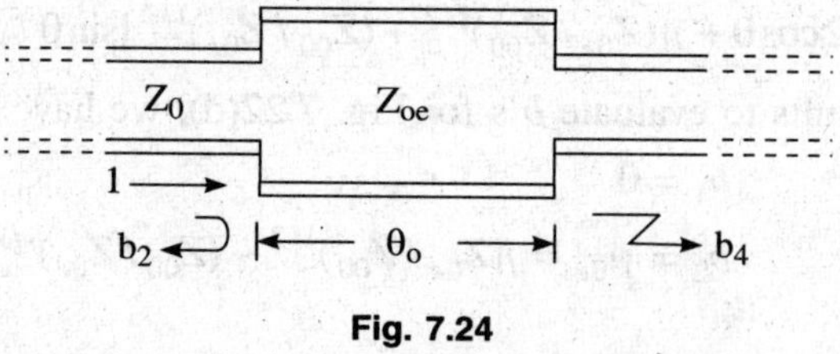

Fig. 7.24

The multiple-section coupler retains the perfect matching and isolation if the even and the odd mode impedances of various sections satisfy the following relation:

$$(Z_{0e1}Z_{001})^{1/2} = (Z_{0e2}Z_{002})^{1/2} = \ldots = (Z_{0en}Z_{00n})^{1/2} = Z_0 \qquad ...(7.103)$$

A three-section coupler is shown in Fig. 7.25. The analysis and design of this coupler is identical to the multi-sectional quarter-wave transformers discussed in Sec. 7.1. The transfer (*ABCD*) matrix for the coupler is obtained by the multiplication of three transfer matrices for the three sections.

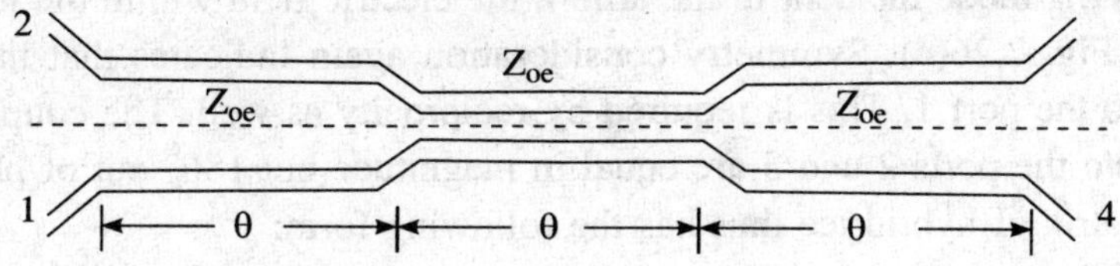

Fig. 7.25

7.3.4 Matched hybrid tee (magic tee)

A matched hybrid tee (sometimes called a magic tee) is an interesting example of a 3 dB waveguide directional coupler. The circuit configuration is shown in Fig. 7.26.

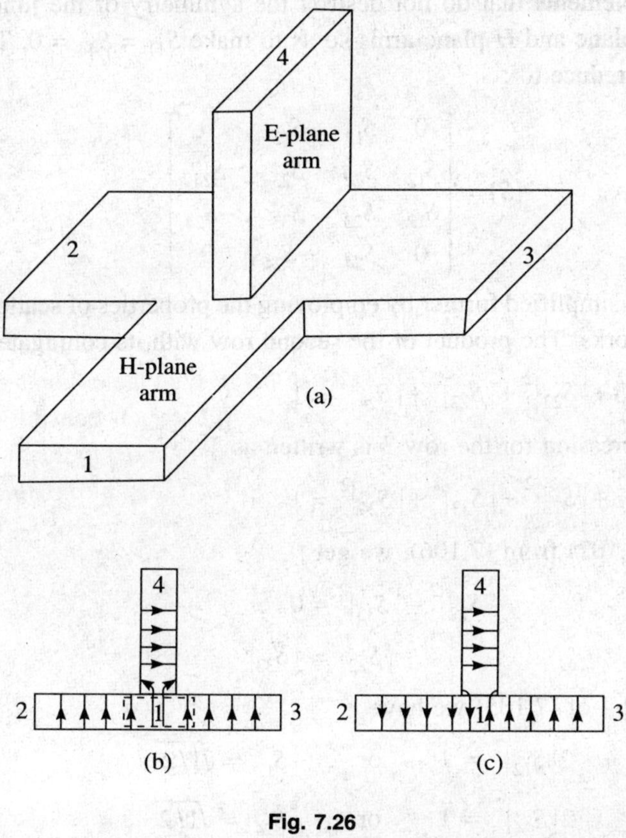

Fig. 7.26

When a TE_{10} mode is incident at the port 1, the electric field within the junction is like that sketched in Fig. 7.26 (b). This electric field has even symmetry about the midplane and hence cannot excite TE_{10} mode in the arm 4 since this mode must have an electric field with odd symmetry. Thus there is no coupling between the ports 1 and 4. The couplings between the ports 1 and 2, and between 1 and 3, are equal, as may be concluded from the symmetry involved.

For a TE_{10} mode incident in the arm 4, the electric field within the junction is sketched in Fig. 7.26(c). Symmetry consideration again indicates that there is no coupling into the port 1. This is required by reciprocity as well. The coupling from the port 4 into the ports 2 and 3 are equal in magnitude but 180° out of phase. The scattering matrix of hybrid tee thus has the following form:

$$[S] = \begin{bmatrix} S_{11} & S_{12} & S_{12} & 0 \\ S_{12} & S_{22} & S_{23} & S_{24} \\ S_{12} & S_{23} & S_{33} & -S_{24} \\ 0 & S_{24} & -S_{24} & S_{44} \end{bmatrix} \quad ...(7.104)$$

Since $S_{12} = S_{13}$, $S_{24} = -S_{34}$ from symmetry.

Matching elements that do not destroy the symmetry of the junction may be placed in the *E*-plane and *H*-plane arms so as to make $S_{11} = S_{44} = 0$. The scattering matrix will then reduce to :

$$[S] = \begin{bmatrix} 0 & S_{12} & S_{12} & 0 \\ S_{12} & S_{22} & S_{23} & S_{24} \\ S_{12} & S_{23} & S_{33} & -S_{24} \\ 0 & S_{24} & -S_{24} & 0 \end{bmatrix} \quad ...(7.105)$$

This can be simplified further by employing the properties of scattering matrices for lossless networks. The product of the second row with its conjugate gives

$$|S_{12}|^2 + |S_{22}|^2 + |S_{23}|^2 + |S_{24}|^2 = 1 \quad ...(7.106)$$

and a similar expression for the row 3 is written as

$$|S_{12}|^2 + |S_{23}|^2 + |S_{33}|^2 + |S_{24}|^2 = 1 \quad ...(7.107)$$

If we subtract (7.107) from (7.106), we get

$$|S_{22}|^2 - |S_{33}|^2 = 0$$

Hence,
$$|S_{22}| = |S_{33}| \quad ...(7.108)$$

From rows 1 and 4 of (7.105) we have

$$2|S_{12}|^2 = 1 \quad \text{or} \quad |S_{12}| = \sqrt{1/2}$$

and
$$2|S_{24}|^2 = 1 \quad \text{or} \quad |S_{24}| = \sqrt{1/2}$$

Thus
$$|S_{12}| = |S_{24}| = \sqrt{1/2} \quad ...(7.109)$$

Using this in equation (7.106) above we have

$$|S_{22}|^2 + |S_{23}|^2 = 0$$

This sum can be zero only if both S_{22} and S_{23} separately become zero. Since $S_{22} = S_{33}$, the coefficient S_{33} should also be zero. The resulting form of the scattering matrix then becomes

$$[S] = \begin{bmatrix} 0 & S_{12} & S_{12} & 0 \\ S_{12} & 0 & 0 & S_{24} \\ S_{12} & 0 & 0 & -S_{24} \\ 0 & S_{24} & -S_{24} & 0 \end{bmatrix} \quad \text{...(7.110)}$$

By proper choice of terminal planes in the arms 1 and 4, we can make S_{12} and S_{24} both real. Thus the scattering matrix of a magic tee can be written in the form

$$S = \sqrt{1/2}\begin{bmatrix} 0 & 1 & 1 & 0 \\ 1 & 0 & 0 & 1 \\ 1 & 0 & 0 & -1 \\ 0 & 1 & -1 & 0 \end{bmatrix} \quad \text{...(7.111)}$$

The matched hybrid tee is used extensively in various applications like balanced mixers.

SELECT BIBLIOGRAPHY

R.E. Collin, ***Foundations for Microwave Engineering,*** McGraw-Hill Book Company, 1966. Chapter 5 on Impedance Transformation and Matching, Chapter 6 on Passive Microwave Devices and Chapter 8 on Periodic Structures and Filters.

G.L. Matthaei, L. Young and E.M.T. Jones, *Microwave Filters, Impedance Matching Networks and Coupling Structures,* McGraw-Hill Book Company, 1964.

R. Levy, 'Directional Couplers', *Advances in Microwaves,* Vol. 1, Academic Press, 1966, p. 186.

H. Howe, *Stripline Circuit Design,* Artech House, 1974.

PROBLEMS

7.1 A 50 ohm transmission line is to be matched to a 100 ohm resistive load. Design a matching transformer using a single short circuited stub of 50 ohm line. What is the value of maximum VSWR over a 20 per cent (± 10 per cent) bandwidth? Also, design a two-section maximally flat transformer and evaluate its bandwidth for an equal value of maximum VSWR.

7.2 Refer to the design of a double stub matching arrangement illustrated in the Fig. 7.3. Show that the load admittances inside the area A may be matched if the stub I is located at a distance $\lambda/4$ from the load.

7.3 Design a two-section Chebyshev transformer to match a line with $Z_0 = 1$ to a load with $Z_L = 2$ with maximum allowable value of $\rho = 0.1$. Calculate Z_1 and Z_2 using eqns. (7.22) and (7.47). Verify the design by calculating ρ at centre frequency.

7.4 Design a three-section binomial (maximally flat) transformer to match a 100 ohm load to a 50 ohm line. The maximum VSWR that can be tolerated is 1.1. What bandwidth can be obtained? Plot ρ versus f.

7.5 Verify the relations (7.13) and (7.14) for the impedance step shown in Fig. 7.6.

7.6 Obtain the scattering matrices for the two networks shown in Fig. 7.15.

7.7 Maximally flat transformer designed in Problem 7.1 is converted into a half wave filter with centre frequency at 3 GHz. Calculate the values of impedance for two half-wave sections of this filter. What is the filter bandwidth if the maximum allowable VSWR is same as in Problem 7.1?

7.8 A three-section maximally flat band-pass filter has a lumped circuit representation shown in Fig. 7.27. Centre frequency is 1.0 GHz, and

$$C_1 = C_3 = 5.1575 \times 10^{-8} \times (2\pi)^{-1}$$

and

$$L_2 = 5.737 \times 10^{-8} \times (2\pi)^{-1}$$

L_1, C_2 and L_3 may be determined from the resonance of the three sections at 1 GHz. Design a transmission line circuit to realize this filter by using quarter-wave-coupled short-circuited shunt stubs. The characteristic impedance of the main line and the terminations at both the ends are 50 ohms each.

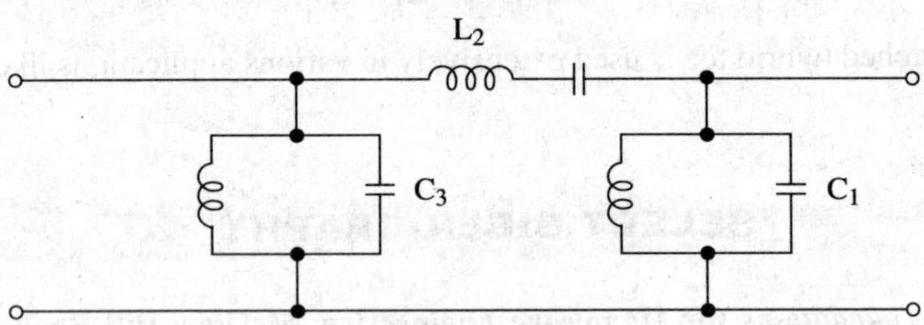

Fig. 7.27

7.9 Derive the relations (7.63) and (7.66) for the input admittance of a short circuited quarter-wave stub and the input impedance of a short circuited half-wave stub, respectively.

7.10 A series L-C resonant circuit is connected in series with one of the conductors of a transmission line. Show that this lumped loading can be represented by a shunt-connected short circuited stub. Find the length and impedance of the stub in terms of L and C.

7.11 A section of a uniform transmission line of length βl is connected at port 1 of a directional coupler whose $[S]$ matrix is given by equation (7.67). It is desired to make $S_{12}(=|S_{12}|\angle\varphi)$ real. Find βl in terms of φ. How will this affect other scattering parameters?

7.12 Two quarter-wave sections of coupled lines are connected in cascade (*see* Fig. 7.25). The even and the odd mode characteristic impedances of the first and the second sections are Z_{0e1}, Z_{001} and Z_{0e1} and Z_{002} respectively. Show that the combination also behaves like a directional coupler when $\sqrt{Z_{0e1}Z_{001}} = \sqrt{Z_{0e2}Z_{002}} = Z_0$, where Z_0 is the characteristic impedance of the lines connected to the four ports of the combination.

7.13 Two 10 dB coupled line directional couplers ($S_{21} = 1/\sqrt{10}$) are connected in cascade as shown in Fig. 7.28. Calculate the characteristics of the composite directional coupler. (The effect of connecting line lengths may be ignored.)

7.14 Two identical couplers are connected in tandem; that is, ports 2 and 4 of the first coupler are connected to ports 1' and 3', respectively of the second coupler.

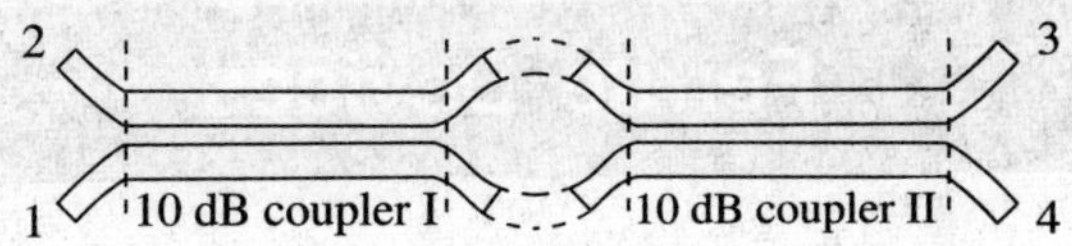

Fig. 7.28

(*see* Fig. 7.29). A cross-over as shown in the figure is needed for tandem connection. Show that the resulting four-port network will also behave as directional coupler.

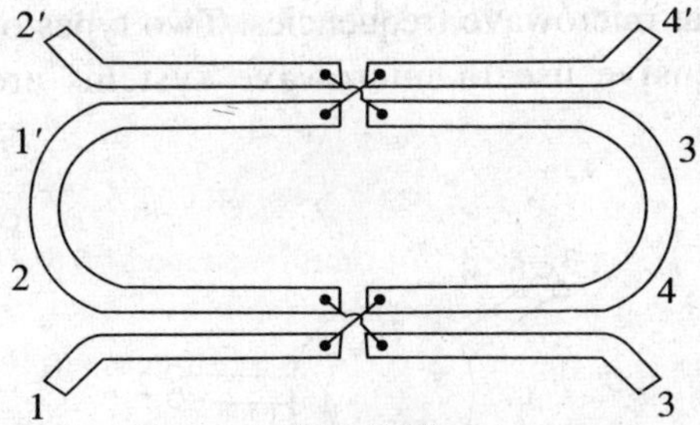

Fig. 7.29

7.15 Consider the tandem combination of couplers discussed in Problem 7.14. If a 3 dB coupling ($S_{21} \equiv 1/\sqrt{2}$) is required from the composite coupler, calculate the coupling of the individual couplers. Discuss the implications of the two possible solutions.

7.16 Two identical impedances Z, twice the value of characteristic impedance of the waveguide, are connected symmetrically to the ports 2 and 3 of a magic tee (Fig. 7.26a). Write the scattering matrix for the two-port network thus formed. If the length of the magic-tee arm at port 2 is made $\lambda/4$ longer than that at port 3, how will the resultant 2-port $[S]$ be modified?

7.17 The arrangement discussed in Problem 7.15 can be used for monitoring the incremental change in the impedance Z at port 3 when the impedance connected to port 2 can be considered to be of the standard value. Derive a relationship between the output at port 4 and an incremental change ΔZ at port 3.

7.18 Consider a series-shunt matching circuit illustrated in Fig. 7.30. This situation results when an *E-H* tuner is used in waveguide circuits. What values of series-stub reactance and shunt-stub susceptance will match an arbitrary load $Y_L = G_L + jB_L$ a distance '*d*' away?

7.19 Develop a graphical method, using the Smith chart to solve Problem 7.18.

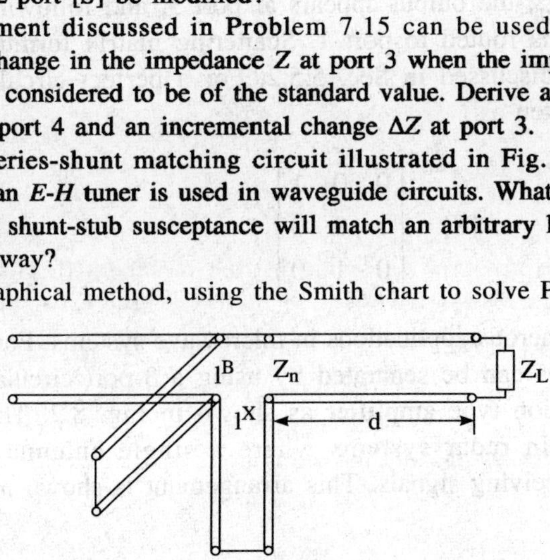

Fig. 7.30

CHAPTER

8

Ferrite Devices

8.1 INTRODUCTION

Characteristics of wave propagation in ferrite materials are used for designing non-reciprocal components at microwave frequencies. Two types of non-reciprocal ferrite devices that find extensive use in microwave systems are: (a) circulators and (b) isolators.

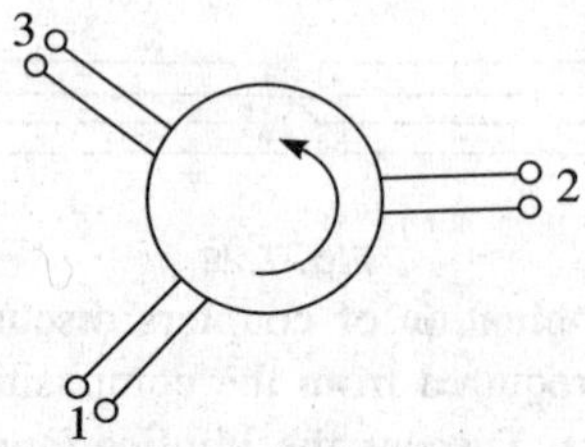

Fig. 8.1

A circulator has three (or more) ports. A 3-port circulator (shown in Fig. 8.1) possesses the following characteristics: (i) all the three ports are matched, and (ii) a signal incident at port 1 appears at port 2 (not at port 3). When there is an input at port 2, the output appears at port 3, and following the cyclic order an input at port 3 is routed to port 1. Scattering matrix formulation for a 3-port network has been discussed in Sec. 5.5.2. For a perfect circulator the scattering matrix may be written as

$$[S] = \begin{bmatrix} 0 & 0 & 1 \\ 1 & 0 & 0 \\ 0 & 1 & 0 \end{bmatrix}$$

Circulators find numerous applications in microwave systems. For example, incident and reflected signals can be separated by using a 3-port circulator. This property is used in a reflection type amplifier as shown in Fig. 8.2. Three-port circulator is also employed in radar systems where a single antenna is used both for transmitting and receiving signals. This arrangement is shown in Fig. 8.3.

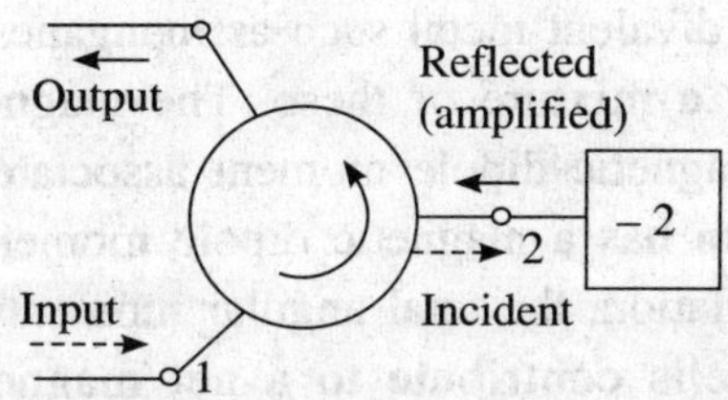

Fig. 8.2

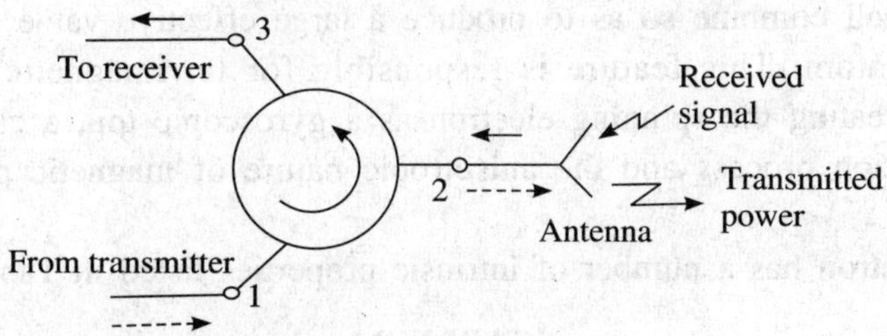

Fig. 8.3

Isolator is a two-port non-reciprocal network. In one direction (say from port 1 to port 2) the signal is transmitted without any attenuation, while in the reverse direction (port 2 to port 1) there is no transmission. Any power incident at port 2 is absorbed inside the isolator. Isolators are very useful components for avoiding the interaction between various parts of a system. Frequency pulling of an RF source by the load can be eliminated by connecting an isolator at the output of the source. With this arrangement, shown in Fig 8.4, the source always looks into a matched load (equal to the characteristic impedance of the isolator).

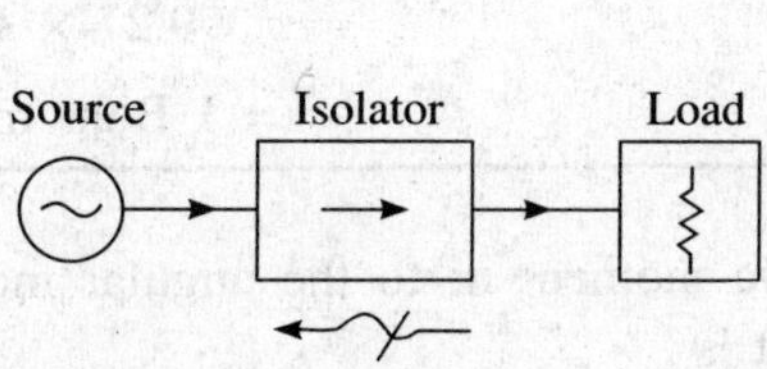

Fig. 8.4

For understanding the working of circulators and isolators it is necessary to study the propagation of electromagnetic waves through a magnetized ferrite medium. This is discussed briefly in the following section.

8.2 FERRITES AND TENSOR PERMEABILITY

Ferrites are ceramic-like materials with a dielectric constant of 10 or more and with specific resistivities that may be as high as 10^{14} times those of metals. Relative permeabilities of several thousand are common. Ferrites are made by sintering a mixture of metallic oxides and have general chemical composition of $MOFe_2O_3$, where M is a divalent metal such as manganese, magnesium, iron, zinc, nickel, cadmium, etc. or a mixture of these. The magnetic properties of ferrites arise mainly from the magnetic dipole moment associated with the electron spin. Every electron in an atom has a magnetic dipole moment associated with it. But, because of mutual cancellation, the total angular momentum of a full shell is zero. Only partially filled shells contribute to a net magnetic dipole moment. For transition elements (such as Cr, Mn, Fe, Co, Ni, etc.) new shells of electrons are formed even before the inner shell is completely filled. Electron spins in the incomplete shell combine so as to produce a large effective value of total dipole moment per atom. This feature is responsible for ferrimagnetic behaviour of ferrites. By treating the spinning electron as a gyroscopic top, a classical picture of magnetization process and the anisotropic nature of magnetic properties may be obtained.

The electron has a number of intrinsic properties listed in Table 8.1.

TABLE 8.1

Intrinsic Properties of an Electron

Property	Notation	Numerical Value	Units
Charge	$-e$	-1.602×10^{-19}	Coulomb
Mass	m_e	9.107×10^{-31}	Kilogram
Angular momentum	$P = \frac{1}{2}\hbar = \frac{1}{2}\frac{h}{2\pi}$	0.527×10^{-34}	Joule-second
moment		9.27×10^{-24} (= 1 Bohr magneton)	Amp-meter^2

The ratio of the magnetic moment **m** to the angular momentum **P** is called the gyromagnetic ratio γ; that is

$$\gamma = \frac{|\mathbf{m}|}{|\mathbf{P}|} = \frac{e}{m_e} \qquad ...(8.1)$$

If an electron is located in a uniform static magnetic field $\mathbf{B}_0$, a torque **T** given by

$$\mathbf{T} = \mathbf{m} \times \mathbf{B}_0 = -\gamma\,[\mathbf{P} \times \mathbf{B}_0] \qquad ...(8.2)$$

will be exerted on the dipole. The negative sign arises because the angular momentum **P** and magnetic dipole moment **m** are anti-parallel for an electron. The torque **T** will cause the dipole axis to precess about an axis parallel to $\mathbf{B}_0$ as shown in Fig. 8.5.

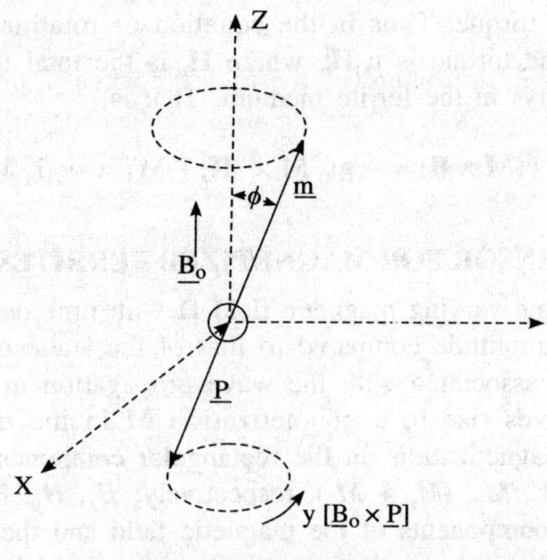

Fig. 8.5

The equation of motion for the electron is obtained from the condition that the rate of change of angular momentum is equal to the torque. Using (8.2), we have

$$\frac{d\mathbf{P}}{dt} = \mathbf{T} = -\gamma\,[\mathbf{P} \times \mathbf{B}_0] = \gamma\,|\,\mathbf{P}\,|\,|\,\mathbf{B}_0\,|\sin\phi \qquad ...(8.3)$$

where ϕ is the angle between **m** and $\mathbf{B}_0$. For free precession, the angular velocity ω_0 is given by

$$\omega_0 = \gamma\,|\,\mathbf{B}_0\,| \qquad ...(8.4)$$

and is independent of the angle φ. The angular velocity ω_0 is often called the *Larmor frequency.* Equation (8.3) may now be written as

$$\frac{d\mathbf{P}}{dt} = \omega_0 \mathbf{P} \sin\phi \qquad ...(8.5)$$

The magnetic properties of a ferrite material may be regarded as due to the collection of N effective spinning electrons per unit volume. Since the spacing between electrons is of atomic dimensions, we may assume the collection of magnetic dipoles as a smeared-out continuous distribution. In the absence of external magnetic field, the dipole moments are oriented at random and the combined effect is very small. When a static magnetic field is applied, the magnetic dipoles get aligned along the direction of the field. This gives the total magnetic dipole moment per unit volume as $\mathbf{M} = N\mathbf{m}$. When the static field $\mathbf{B}_0$ is large enough to cause saturation of the magnetization in the ferrite, $\mathbf{M} = \mathbf{M}_s$. In a saturated ferrite all the spins are so tightly coupled that the whole sample acts essentially as a large single magnetic dipole.

The magnetization $\mathbf{M}_s$ produces a contribution to the total internal **B** field according to the relation $\mathbf{B} = \mu_0\,(\mathbf{H}_0 + \mathbf{M}_s)$. The torque acting on $\mathbf{M}_s$ is due only to the field $\mathbf{B}_0 = \mu_0\mathbf{H}_0$. Since the cross product of $\mu_0\mathbf{M}_s$ with $\mathbf{M}_s$ is zero, it does not contribute to the torque. Thus in the equation of rotation of magnetization, the field producing the torque is $\mu_0\mathbf{H}_t$, where $\mathbf{H}_t$ is the total (static plus any AC magnetic field intensity) in the ferrite medium. That is,

$$\frac{d\mathbf{M}}{dt} = -\gamma\,(\mathbf{M} \times \mathbf{B}) = -\gamma\mu_0\,\mathbf{M} \times (\mathbf{H}_t + \mathbf{M}) = -\gamma\mu_0\,\mathbf{M} \times \mathbf{H}_t \qquad \text{...(8.6)}$$

PERMEABILITY TENSOR FOR MAGNETIZED FERRITES

Let there also be a time varying magnetic field $\mathbf{H}$ with time dependence exp $(j\omega t)$ and having a small amplitude compared to that of the static magnetic field. The time varying field is associated with the wave propagation in the medium. This magnetic field $\mathbf{H}$ gives rise to a magnetization $\mathbf{M}$ in the material. The total magnetic field and magnetization (in the rectangular components) consists of H_x, H_y, $(H_z + H_0)$ and M_x, M_y, $(M_z + M_0)$, respectively; H_x, H_y, H_z and M_x, M_y, M_z are the time-varying components of the magnetic field and the magnetization. H_0 and M_0 are static field and the static magnetization, respectively. Substituting these field components in (8.6), we have

$$\begin{aligned} j\omega M_x &= -\gamma\mu_0 M_y(H_0 + H_z) + \gamma\mu_0(M_0 + M_z)H_y \\ j\omega M_y &= -\gamma\mu_0(M_0 + M_z)H_x + \gamma\mu_0 M_x(H_0 + H_z) \\ j\omega M_z &= -\gamma\mu_0(M_x H_y) + \gamma\mu_0 M_y H_x \end{aligned} \qquad \text{...(8.7)}$$

If it is assumed that the time-varying magnetic fields and the corresponding magnetizations are much smaller than the static fields, all the terms containing the products of the components of the time-varying quantities may be neglected. Equations (8.7) can now be written as

$$\begin{aligned} j\omega M_x + \gamma\mu_0 H_0 M_y &= \gamma\mu_0 M_0 H_y \\ -\gamma\mu_0 H_0 M_x + j\omega M_y &= -\gamma\mu_0 M_0 H_x \\ j\omega M_z &= 0 \end{aligned} \qquad \text{...(8.8)}$$

It will be seen that the first two equations in (8.8) are simultaneous equations in M_x and M_y. They may be easily solved. The following notations are adopted

$$\kappa = \frac{-\omega\gamma M_0\mu_0}{\mu_0^2\gamma^2 H_0^2 - \omega^2} = \frac{\omega\omega_m}{\omega_0^2 - \omega^2} \qquad \text{...(8.9)}$$

$$\psi = \frac{\gamma^2 H_0 M_0 \mu_0^2}{\mu_0^2\gamma H_0^2 - \omega^2} = \frac{\omega_0\omega_m}{\omega_0^2 - \omega^2} \qquad \text{...(8.10)}$$

where $\omega_0 = \gamma\mu_0 H_0$ and $\omega_m = \gamma\mu_0 M_0$.

Equation (8.8) may be rewritten as

$$\begin{aligned} M_x &= \psi H_x - j\kappa H_y \\ M_y &= j\kappa H_x + \psi H_y \\ M_z &= 0 \end{aligned} \qquad \text{...(8.11)}$$

The total magnetic flux density is $\mathbf{B} = \mu_0 (\mathbf{H} + \mathbf{M})$. Components of **B** may therefore be written as

$$B_x = \mu_0[(1+\psi)H_x - j\kappa H_y] \qquad ...(8.12)$$

$$B_y = \mu_0[(1+\psi)H_y + j\kappa H_x] \qquad ...(8.13)$$

$$B_z = \mu_0 (H_z + H_0) \qquad ...(8.14)$$

The relationship between **B** and **H** can also be expressed as

$$\begin{bmatrix} B_x \\ B_y \\ B_z \end{bmatrix} = \mu_0 \begin{bmatrix} (1+\psi) & -j\kappa & 0 \\ j\kappa & (1+\psi) & 0 \\ 0 & 0 & 1 \end{bmatrix} \begin{bmatrix} H_x \\ H_y \\ H_z + H_0 \end{bmatrix} \qquad ...(8.15)$$

This matrix representation of the relationship between two vectors **B** and **H** defines the *tensor permeability* for a magnetized ferrite material.

8.3 WAVE PROPAGATION IN A FERRITE MEDIUM

In this section we shall consider the propagation of a plane electromagnetic wave through a ferrite medium immersed in a static magnetic field $\mathbf{H}_0$. The direction of propagation is taken to be along the magnetic field. Effective permeability is given by (8.15) which corresponds to a relation between magnetic flux **B** and magnetic field **H**. This relationship may be rewritten as

$$B_x = \mu_1 H_x - j\mu_2 H_y$$

$$B_y = j\mu_2 H_x + \mu_1 H_y$$

$$B_z = \mu_0 (H_z + H_0)$$

where $\mu_1 = 1 + \psi$ and $\mu_2 = \kappa$.

Maxwell's equation for curl **E** may now be written as

$$\frac{\partial E_z}{\partial y} - \frac{\partial E_y}{\partial z} = -j\omega\mu_1 H_x - \omega\mu_2 H_y$$

$$\frac{\partial}{\partial z}E_x - \frac{\partial}{\partial x}E_z = \omega\mu_2 H_x - j\omega\mu_1 H_y$$

$$\frac{\partial E_y}{\partial x} - \frac{\partial E_x}{\partial y} = -j\omega\mu_0 H_z \qquad ...(8.16)$$

Since we are considering a plane wave propagation along z-direction

$$\frac{\partial}{\partial x} = \frac{\partial}{\partial y} = 0 \qquad ...(8.17)$$

Taking the field variation as exp $(j\omega t - j\beta z)$ we can rewrite (8.16) as

$$\beta E_y = -\omega\mu_1 H_x + j\omega\mu_2 H_y$$

$$\beta E_x = j\omega\mu_2 H_x + \omega\mu_1 H_y$$

and

$$0 = H_z \qquad ...(8.18)$$

Similarly, from Maxwell's equation for Curl **H** we get

$$\beta H_y = \omega\varepsilon E_x$$

$$\beta H_x = -\omega\varepsilon E_y$$

$$0 = E_z \qquad ...(8.19)$$

By eliminating H_x and H_y from equations (8.18) and (8.19) we obtain the following two relationships between E_x and E_y

$$(\beta^2 - \omega^2\mu_1\varepsilon)E_y = j\omega^2\varepsilon\mu_2 E_x \qquad ...(8.20)$$

$$(\beta^2 - \omega^2\mu_1\varepsilon)E_x = -j\omega^2\varepsilon\mu_2 E_y \qquad ...(8.21)$$

For both of these relations to be satisfied simultaneously, we should have

$$(\beta^2 - \omega^2\mu_1\varepsilon)^2 = \omega^4\varepsilon^2\mu_2^2$$

or

$$\beta^2 = \omega^2\varepsilon(\mu_1 \pm \mu_2) \qquad ...(8.22)$$

This indicates that two distinct values of β^2 are possible. The corresponding expressions for phase constant β are :

$$\beta^+ = \pm\omega\sqrt{\varepsilon(\mu_1 \pm \mu_2)}$$

and

$$\beta^- = \pm\omega\sqrt{\varepsilon(\mu_1 - \mu_2)} \qquad ...(8.23)$$

For both the modes, characterized by values of β given by (8.23), the relationship between E_x and E_y is given by (8.20), i.e.

$$E_x = \frac{\beta^2 - \omega^2\mu_1\varepsilon}{j\omega^2\varepsilon\mu_2}E_y \qquad ...(8.24)$$

When $\beta = \beta^+ = \omega\sqrt{\varepsilon_1(\mu_1 + \mu_2)}$, we get $E_x = -j\,E_y$ whereas when $\beta = \beta^-$, $E_x = +j\,E_y$. Thus, β^+ and β^- correspond to two circularly polarized plane waves. For $\beta = \beta^+$, E_x lags E_y and **E** phasor moves in clockwise direction. For $\beta = \beta^-$, **E** phasor moves in counter-clockwise direction.

These two circularly polarized waves may be considered as fundamental modes of propagation in a ferrite medium. Since a linearly polarized wave can be considered as composed of two circularly polarized waves any plane wave propagating in the magnetized ferrite medium may be analysed in terms of the two fundamental modes characterized by β^+ and β^-.

Effective permeability for the two modes characterized by β^+ and β^- are denoted by μ^+ and μ^- and are given by

$$\mu^+ = \mu_1 + \mu_2 = \mu_0[1 + \omega_m/(\omega_0 - \omega)] \qquad ...(8.25a)$$

$$\mu^- = \mu_1 - \mu_2 = \mu_0[1 + \omega_m/(\omega_0 + \omega)] \qquad ...(8.25b)$$

Variations of μ^+ and μ^- with frequency are presented in Fig. 8.6. It may be noted that μ^+ becomes infinite and changes sign at $\omega = \omega_0$ (gyromagnetic resonance frequency). On the other hand, μ^- varies monotonically from $\mu_0(1 + \omega_m/\omega_0)$ to μ_0 as the frequency is varied from zero to infinite. Thus only one of the modes (characterized by μ^+) exhibits resonance behaviour at $\omega = \omega_0$.

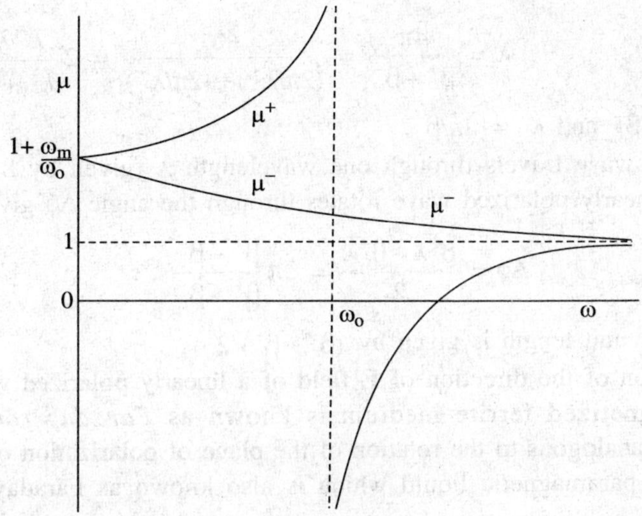

Fig. 8.6

Existence of different propagation characteristics for two circularly polarized waves having opposite sense of rotation of E phasors is a significant feature of wave propagation through ferrites.

8.4 FARADAY ROTATION IN FERRITES

Consider a linearly polarized plane wave travelling through a magnetized ferrite medium. As discussed in the previous section the linearly polarized wave may be resolved into two circularly polarized waves with opposite senses of rotation. Since these circularly polarized waves have different values of propagation constants (β^+ and β^-), the two electric field (**E**) vectors rotate at different rates. The **E** field vectors of these circularly polarized waves combine to produce a maximum (in almost same direction) when the sum of phase shifts experienced by the two waves is 4π radians.

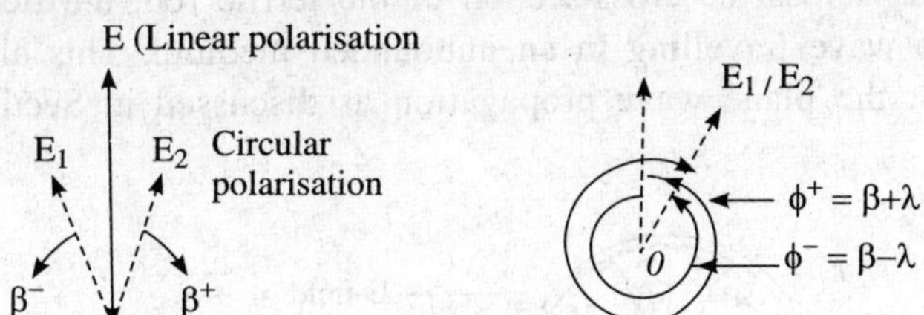

Fig. 8.7

The corresponding distance defines the wavelength λ for the linearly polarized wave. This is illustrated in Fig. 8.7. We may write

$$\beta^+ \lambda + \beta^- \lambda = 4\pi \quad \text{or} \quad \beta^+ + \beta^- = \frac{4\pi}{\lambda} \qquad ...(8.26)$$

$$\lambda = \frac{4\pi}{\beta^+ + \beta^-} = \frac{4\pi}{(2\pi/\lambda^+) + (2\pi/\lambda^-)} = 2\frac{\lambda^-\lambda^+}{\lambda^+ + \lambda^-} \quad ...(8.27)$$

where $\lambda^+ = 2\pi/\beta^+$ and $\lambda^- = 2\pi/\beta^-$.

When the wave travels through one wavelength λ (given by 8.27), **E** field vector of the linearly polarized wave rotates through the angle $\Delta\phi$ given by

$$\Delta\phi = \frac{\beta^+\lambda - \beta^-\lambda}{2} = 2\pi\frac{\beta^+ - \beta^-}{\beta^+ + \beta^-} \quad ...(8.28)$$

The rotation per unit length is given by $(\beta^+ - \beta^-)/2$.

The rotation of the direction of **E** field of a linearly polarized wave passing through a magnetized ferrite medium is known as ***Faraday rotation.*** This phenomenon is analogous to the rotation of the plane of polarization of light when passed through paramagnetic liquid which is also known as Faraday rotation in optics.

Property of Faraday rotation in ferrites is used in designing non-reciprocal devices like isolators and circulators. Working of these devices is discussed in the following sections.

8.5 ISOLATOR

As mentioned earlier, an isolator allows unattenuated transmission of power in one direction whereas the power incident from other direction is completely absorbed inside the isolator.

The isolator design discussed here uses a Faraday rotation of the E field vector by 45°. For achieving this rotation, a circular waveguide is used and a cylindrical ferrite rod is placed along the axis of the circular waveguide. The ferrite rod is magnetized with a static magnetic field along the axis of the waveguide which is excited in dominant TE_{11} mode. The field configuration for this mode is shown in Fig. 8.8. We note that **E** and **H** field lines for this mode are almost rectilinear near the axis. Thus across the cross-section of the ferrite rod, the field is similar to that of a plane wave travelling in an unbounded medium. This allows one to use the results for the plane wave propagation as discussed in Sections 8.3 and 8.4.

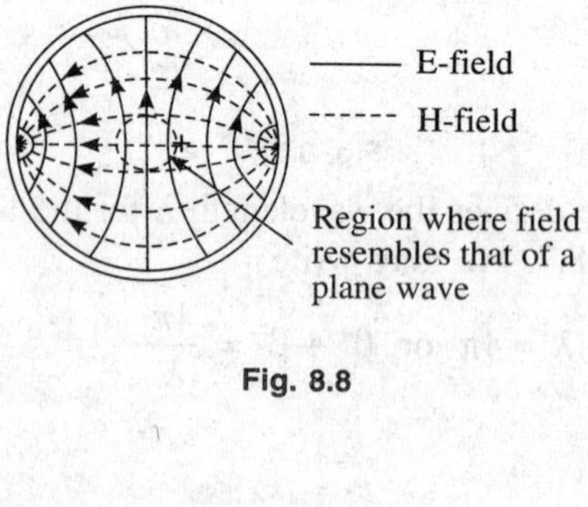

Fig. 8.8

A schematic arrangement for an isolator using a Faraday rotation of 45° is shown in Fig. 8.9. The main part of the isolator is built in a circular waveguide. However, at both the input and the output ends the circular waveguide is tapered to rectangular waveguide sections. As shown in the figure the orientations of the input and output rectangular guides are at 45° to each other. A ferrite rod is placed along the axis in the circular waveguide region of the assembly. The length of the ferrite rod and the magnitude of the static magnetic field is selected such that E field vector of the wave (at the desired operating frequency) travelling along the ferrite (in either direction) gets rotated by 45°. In addition to the ferrite rod, two resistive vanes are also placed inside the waveguide. One of the resistive vanes is located near the input port and the other near the output port. The resistive vanes are placed parallel to the broad walls of the rectangular sections. Edges of the vanes looking towards the ferrite rod are tapered as shown.

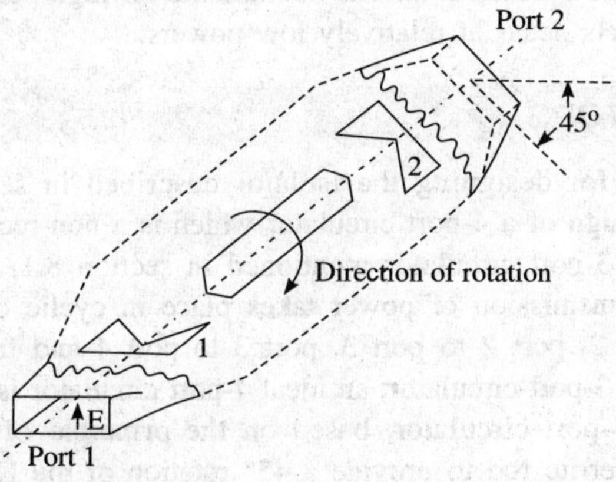

Fig. 8.9

The operation of this isolator may be described as follows. A wave incident at port 1 has fixed **E** field direction because of the orientation of the rectangular waveguide section at the input. This excites a TE_{11} mode in circular waveguide. **E** field of this TE_{11} mode is parallel to the **E** field of the incident wave and thus perpendicular to the resistive vane numbered 1 (near the input end). Therefore, the power incident at the port 1 is not affected by the resistive vane near this end. This wave now encounters the magnetized ferrite rod, and as it travels along the rod the direction of the **E** field gets rotated by 45° clockwise (looking into the direction of propagation). After this rotation, the **E** field vector becomes perpendicular to the vane 2 and therefore unaffected by its presence. Since the orientation of port 2 is fixed at an angle of 45° clockwise to port 1, the wave couples to the rectangular waveguide section at the output. Thus the power is transmitted unattenuated from the port 1 to the port 2. Now let us consider a wave incident at the port 2. In this case **E** field direction is such that this wave is unaffected by resistive vane numbered 2. While travelling along the ferrite rod it encounters a static magnetic field in the reverse (to that for the wave from 1 to 2) direction. Therefore, **E** vector rotates counter-clockwise (now looking into the direction of propagation from 2 to 1). The rotation is again in the direction marked in the figure. However, after this rotation, **E** field lines become parallel to the resistive vane numbered 1 and the power is absorbed by this vane. The edge of the vane is tapered to reduce any reflection towards the port 2. Even if such a reflection does take place, the E field of the reflected wave becomes parallel to resistive vane numbered 2 also, and the power gets dissipated in this vane. Thus the power travelling from port 2 towards port 1 is not transmitted but gets totally absorbed inside the isolator.

The ferrite rod and the resistive vanes are supported mechanically by a low loss dielectric material. This is not shown in the figure. The static magnetic field is obtained by a permanent magnet or a solenoid placed outside the waveguide. Practical isolators of this type have forward loss up to 1 dB and reverse isolation upward of 20 dB. The Faraday rotator is not suitable for high peak power operation since nonlinear effects occur at relatively low powers.

8.6 CIRCULATORS

Methodology used for designing the isolator described in section 8.5 may be extended for the design of a 4-port circulator which is a non-reciprocal component very similar to the 3-port circulator mentioned in section 8.1. All the four ports are matched and transmission of power takes place in cyclic order only, that is, from port 1 to port 2, port 2 to port 3, port 3 to port 4 and from port 4 to port 1. As in the case of 3-port circulator, an ideal 4-port circulator is perfectly lossless.

Design of a 4-port circulator, based on the principle of Faraday rotation, uses a magnetized ferrite rod to provide a 45° rotation of the **E** field by using an arrangement identical to that in the case of isolators. Basic configuration is similar to that shown in Fig. 8.9 for isolators. Resistive vanes are no longer required. Instead, arrangements are made to couple out signals that otherwise would have dissipated in resistive vanes in the case of isolators. This is achieved by adding two more ports. These ports are oriented radially outwards and are attached to the circular waveguide section. Angular positions of these two ports (number 3 and 4) with respect to ports 1 and 2 arc shown in Fig. 8.10(a). As shown there, ports 1, 2, 3 and 4 are oriented such that **E** field can couple to these ports successively in numerical order after going through a rotation of 45° in clockwise direction in each step. It may be noted that ports 1 and 3 are on one end of the ferrite rod whereas ports 2 and 4 are on the other end as illustrated in Fig. 8.10(b).

Once the geometrical configuration is identified, the operation of the circulator can be easily understood. For a signal incident at port 1, **E** vector is in the direction '*m*' (*see* Fig. 8.10). When the wave travels along the magnetized ferrite, the direction of the **E** field vector gets rotated by 45° to coincide with the direction '*n*'. Thus the power incident at the port 1, appears at the port 2. It does not get coupled to the port 4, since ports 2 and 4 are at 90° to each other. By similar arguments one can see that signal incident at port 2 will appear at port 3 and so on.

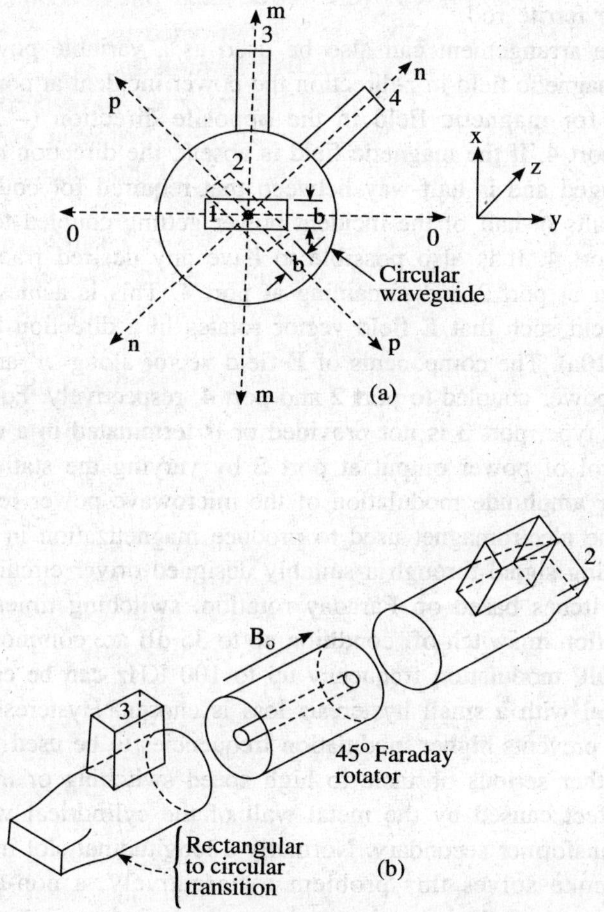

Fig. 8.10

It may be noted that this scheme cannot be easily adopted for a 3-port circulator. However, any 4-port circulator can be converted into a 3-port circulator by terminating one of the ports in a perfect short circuit. The most commonly used design of a 3-port circulator is, however, based on a different principle and is discussed, in Sec. 8.8.

8.7 FARADAY ROTATION SWITCH AND MODULATOR

The 4-port circulator described in the previous section can be easily converted into a single-pole-double-throw (SPDT) switch. Switching action is obtained by reversing the direction of magnetization. This changes the rotation of **E** field vector from the clockwise to the anti-clockwise direction and couples the power to the port 4 instead of the port 2 as was the case earlier. Thus port 1 can be connected to port 2 or port 4 by changing the direction of current in the electromagnet used for magnetizing the ferrite rod.

The above arrangement can also be used as a variable power divider. We recall that for magnetic field in z-direction the power incident at port 1 gets coupled to port 2, and for magnetic field in the opposite direction ($-z$) the power is transmitted to port 4. If the magnetic field is absent, the direction of **E** field vector remains unchanged and is half-way between that required for coupling to ports 2 and 4. This results in half of the incident power getting coupled to port 2 and the other half to port 4. It is also possible to have any desired fraction of incident power to appear at port 2 and remaining at port 4. This is achieved by adjusting the magnetic field such that **E** field vector rotates in a direction between '*n*' and '*p*' (*see* Fig. 8.10a). The components of **E** field vector along '*n*' and '*p*' determine the amount of power coupled to port 2 and port 4, respectively. For variable power dividers of this type, port 3 is not provided or is terminated in a matched load.

The control of power output at port 2 by varying the static magnetic field can be used for amplitude modulation of the microwave power fed to port 1. For this purpose, the electromagnet used to produce magnetization in ferrite is driven by the modulating signal through a suitably designed driver circuit.

In the switches based on Faraday rotation, switching times of 0.1 to 0.01 sec and attenuation in switch-off condition up to 35 dB are common. For amplitude modulator circuit, modulation frequency up to 100 KHz can be easily achieved if a ferrite material with a small hysteresis loss is chosen. Hysteresis heating of the ferrite material prevents higher modulation frequencies to be used unless the ferrite is cooled. Another serious obstacle to high speed switching or modulation is the shorted-turn-effect caused by the metal wall of the cylindrical waveguide acting as a shorted transformer secondary. Normally a longitudinal slot in the guide fitted with an RF choke solves this problem. Alternatively, a non-conducting tube supporting a thin metal film may be used as a waveguide.

8.8 3-PORT CIRCULATORS

The most frequently used ferrite component is a 3-port junction circulator in waveguide or stripline configurations as shown in Figs. 8.11 and 8.12, respectively. In the waveguide configuration, we have a cylindrical disc of ferrite placed symmetrically with respect to the three ports as shown. In stripline configuration there are two ferrite cylinders, one on either side of the strips. A magnetizing field is applied parallel to the axis of the ferrite cylinders. A phenomenological description of this type of junction circulator is presented in the following paragraphs.

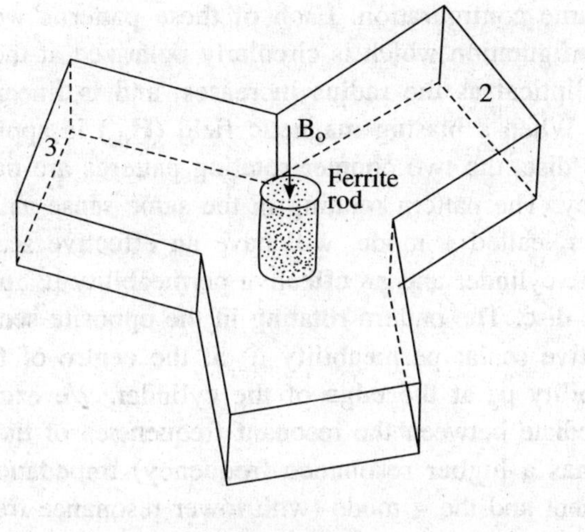

Fig. 8.11

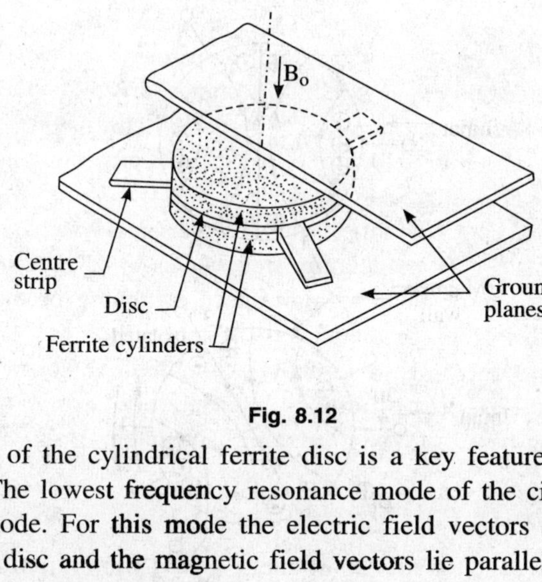

Fig. 8.12

Resonance of the cylindrical ferrite disc is a key feature in the operation of this circulator. The lowest frequency resonance mode of the circular disc structure is the dipolar mode. For this mode the electric field vectors are perpendicular to the plane of the disc and the magnetic field vectors lie parallel to the plane of the disc. This mode configuration when excited from port 1, is illustrated in Fig. 8.13(a), for the unmagnetized case. For stripline configuration the lower ferrite disc behaves as a mirror image of the top one. The present discussion is in terms of a single disc only. Referring to the field configuration shown in Fig. 8.13(a), we note that the electric field components at ports 2 and 3 are equal. If this field pattern is rotated, as shown in Fig. 8.13(b), then the port 3 is situated at the null of the electric field and electric field components at ports 1 and 2 are equal in magnitude. The device is now equivalent to a transmission cavity between ports 1 and 2 and the port 3 is isolated.

The field pattern of Fig. 8.13(b) can be generated by two counter rotating patterns of the same configuration. Each of these patterns would involve an RF magnetic field configuration which is circularly polarized at the centre of the disc, becomes more elliptical as the radius increases, and is linearly polarized at the edge of the disc. When a biasing magnetic field (H_{dc}) is applied in the direction of the axis of the disc, the two counter rotating patterns are no longer resonant at the same frequency. The pattern rotating in the same sense as the electron spin of the ferrite cylinder, called + mode, will have an effective scalar permeability μ^+ at the centre of the cylinder and an effective permeability μ_e for linear polarization at the edge of the disc. The pattern rotating in the opposite sense is called – mode and has an effective scalar permeability μ^- at the centre of the cylinder and an effective permeability μ_e at the edge of the cylinder. We excite the system at a frequency intermediate between the resonant frequencies of these two modes. The + mode (which has a higher resonance frequency) impedance has an inductive reactance component and the – mode (with lower resonance frequency) impedance has a capacitive reactance component. If the frequency is chosen such that the

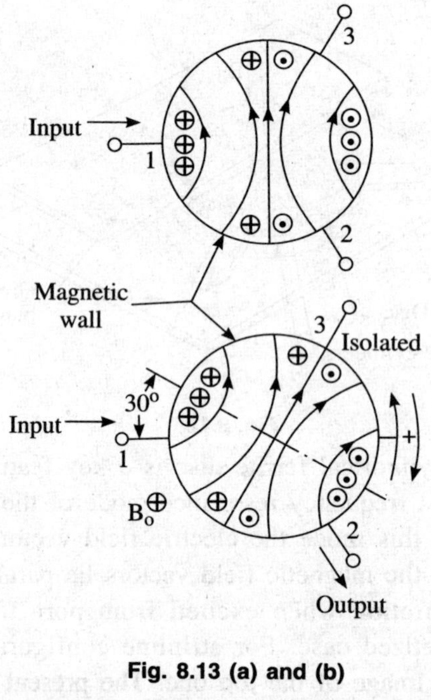

Fig. 8.13 (a) and (b)

capacitive reactance component of one equals the inductive reactance component of the other, the total impedance is real. When the degree of splitting is adjusted such that the phase angles of the impedances of the two modes are each 30° at the operating frequency, the standing wave pattern will be rotated to 30° from the one in unmagnetized case. The + mode which has the inductive reactance component will have its voltage-maximum leading the current (or **H** field) maximum at the input port by 30° in time phase, which is also 30° in space phase for this mode. The – mode will have its voltage-maximum lagging the current-maximum at the input port by 30°. Therefore, the two voltage maxima will coincide at a point 30° away from the input port as shown in Fig. 8.13 (b). This 30° rotation brings the **E** field null of the pattern to the port 3 so that no voltage exists at this port. However, for the power entering at the port 1, the device acts as a transmission cavity with power leaving at the port 2. Since the device has complete symmetry, any power entering at port 2 will set up a new set of similar modes resulting in this power leaving at port 3, and port 1 will be isolated. Similarly, power entering at port 3 is transmitted to port 1 and port 2 is isolated.

This type of circulator is a very versatile component at microwave frequency. In addition to its use as a circulator, it can also be used as an isolator (by terminating one of the ports in a matched load) or as a switch (by reversing $\mathbf{H}_{dc}$).

8.9 RESONANCE ABSORPTION IN FERRITES

It is seen from equation (8.25a) that as the operating frequency approaches the cyclotron resonance frequency ω_0, the effective pecmeability for a circularly polarized wave (μ^+) tends to infinity. This makes the phase constant β^+ also tend to infinity and the phase velocity of the wave tends to zero. However, in the formulation used in this analysis, the damping in the ferrite medium has not been taken into account. A more complete analysis shows that the damping represented by the imaginary part of complex permeability has a sharply peaked maximum at the resonance frequency for a circularly polarized (counter-clockwise) wave. A circularly polarized wave with the field vector rotating in the opposite direction has an effective permeability of μ^- and hence does not go through absorptive resonance at ω equal to ω_0.

This non-reciprocal absorption property of ferrite (at cyclotron resonance frequency) can be used to design an isolator. In this type of isolator the power flowing in the reverse direction is absorbed in the ferrite material itself and any other absorbing material need not be included. If a circularly polarized wave at ferrimagnetic resonance frequency propagates in a magnetized ferrite material, in one of the directions the effective permeability is μ^+ and absorption takes place. However, if the direction of propagation is reversed, the wave sees an effective permeability μ^- and is transmitted without any attenuation. In order to fabricate a practical device based on this principle, it is necessary that the ferrite material be placed inside the waveguide at such a location that it sees a circularly polarized field.

On studying the field distribution for TE_{10} mode inside a rectangular waveguide, it is found that : (i) **E** field has only one component and cannot therefore be circularly polarized at any location, (ii) magnetic field has two normal components which are 90° out of phase. Thus at any point where these two components are equal in magnitude we have a circularly polarized magnetic field. From the expressions for two magnetic field components (equation 2.71), we find that the above condition is satisfied when the distance from the narrow wall (x_c in Fig. 8.14) is given by

$$x_c = (a/\pi)\tan^{-1}(\pi/\beta a) \quad ...(8.29)$$

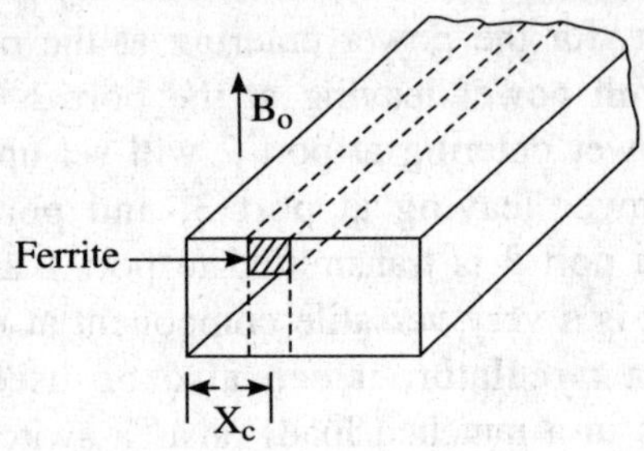

Fig. 8.14

where 'a' is the wide dimension of the waveguide and β is the propagation constant. Taking advantage of this property, the ferrite material is located as shown in Fig. 8.14. The direction of the static magnetic field required is also shown in this figure. As illustrated in Fig. 8.15, for a wave travelling to the right the **H** field vector at point P moves in counter-clockwise direction, whereas for a wave moving towards left the H field vector rotates in clockwise direction.

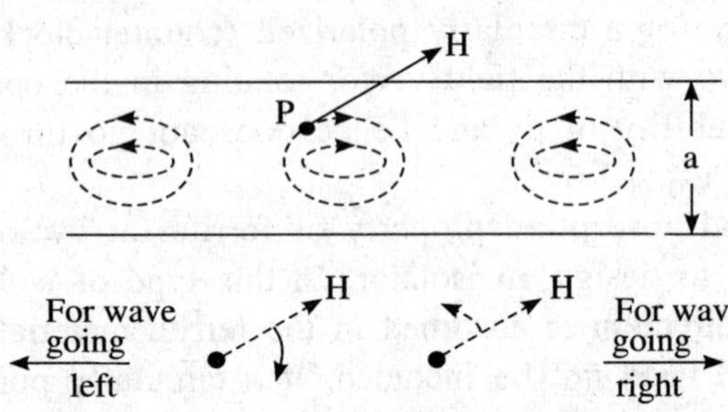

Fig. 8.15

An advantage of the design shown in Fig. 8.14 is that the ferrite material is in thermal contact with the metal wall of the guide, which acts as a heat-sink for the heat generated within ferrite by the resonance absorption process.

A resonance isolator can also be constructed in a circular, coaxial ridged or stripline waveguide or any other transmission structure in which it is possible to find or to create a condition such that, in some portion of the guide, the microwave field vector rotates as the wave pattern moves along the guide.

8.10 YIG RESONATORS

Yttrium iron garnet (YIG) is one of the rare-earth-iron-garnets identified by the chemical formula $Me_3Fe_5O_{12}$ or $5Fe_2O_33Me_2O_3$ or $Me_3Fe_2(FeO_4)_3$ where Me denotes a rare-earth-metal. These garnets show ferrimagnetic properties similar to ordinary ferrites but exhibit much smaller losses. Losses in garnets become exceptionally low when single crystals in the form of small spheres (typically 40 mil diameter) are used. Because of very low losses the garnet spheres can be used as resonators with the frequency of ferrimagnetic resonance given by

$$\omega_0 = \mu_0 \gamma H_0 \qquad \text{...(8.30a)}$$

which may be written as

$$f_0 = 2.8\, H_0 \qquad \text{...(8.30b)}$$

where f_0 is in megaHertz and $\mathbf{H}_0$ in oersteds.

An important characteristic of the garnet resonators is that the resonance frequency may be varied by changing $\mathbf{H}_0$. Thus one has a resonator whose frequency may be varied electronically over a wide range. Lower limit arises because of the minimum $\mathbf{H}_0$ required for saturation magnetization. For pure YIG this limit is around 3.5 GHz. However, when YIG is doped with Gadolinium (Gd), the value of saturation magnetization can be reduced.

Coupling to a YIG resonator is through the magnetic field of the incident electromagnetic wave. A typical arrangement using loops of electric current is shown in Fig. 8.16. This arrangement consists of a YIG sphere at the centre of two loops whose axes are perpendicular to each other and also to the DC field H_0.

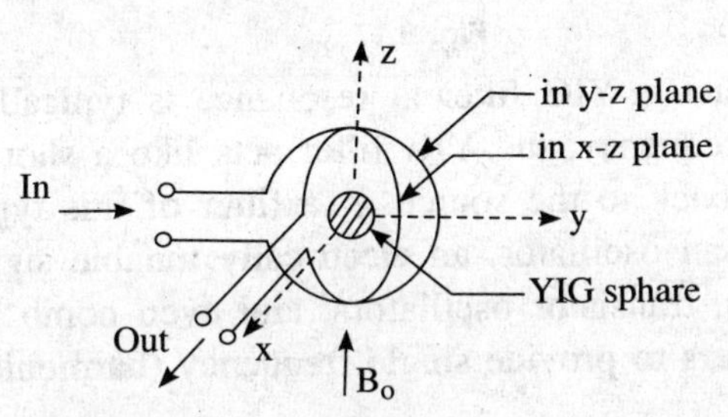

Fig. 8.16

One of the loops carries the input RF current, while the other is connected to the load. When $\mathbf{H}_0$ is zero, there is large input-output isolation since the two loops are perpendicular. With $\mathbf{H}_0$ applied, there is a net magnetization vector in the direction of $\mathbf{H}_0$. The magnetic field component H_x produced by the RF driving current in the input loop causes the net magnetization vector to precess about the z-axis. This mechanism can alternatively be visualized through the relation $d\mathbf{M}/dt = -\gamma\mu_0 (\mathbf{M} \times \mathbf{H})$, which indicates that a y-component of magnetization will be produced. The resulting RF magnetization component (M_y) induces a voltage in the output loop. At frequencies away from the ferrimagnetic resonance frequency, M_y and the voltage it induces are small, and therefore the input-to-output isolation is high. When the input current is at the ferrimagnetic resonance frequency, M_y is maximum. At this frequency, there is a large transfer of power from input to output and the insertion loss is low. Thus the arrangement shown in Fig. 8.16 behaves like a filter whose centre frequency is the ferrimagnetic resonance frequency and which can be tuned by varying $\mathbf{H}_0$. The YIG sphere and the RF structure are located in the field of an electromagnet and the tuning is accomplished by changing the magnet current.

A YIG filter with increased off-band isolation can be designed by using two YIG spheres as shown in Fig. 8.17. Two spheres are placed in the path of the RF signal. The signal can pass through the YIG filter only when the spheres are at resonance.

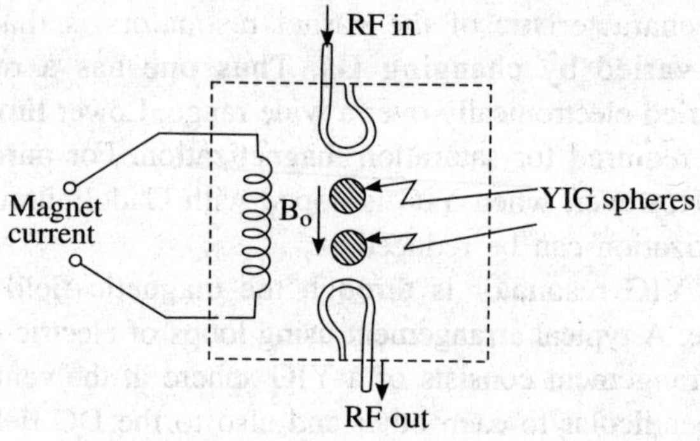

Fig. 8.17

Insertion loss of such a YIG filter at resonance is typically less than 5 dB. Away from resonance frequency, the YIG filter acts like a short circuit reflecting most of the RF energy back to the source. If a filter of this type is connected in the feed-back circuit of an oscillator, an electrically tunable signal source can be designed. Tunnel diodes, transistor oscillators, and even comb generators can be connected with these filters to provide single frequency (harmonic free) microwave signal sources.

A waveguide version of YIG tuned filter is shown in Fig. 8.18. The coupling arrangement used here is analogous to the crossed loops system of Fig. 8.16. In the waveguide version the magnetic fields of the dominant modes in the two waveguides are perpendicular to each other at the location of YIG sphere. The single crystal YIG sphere is mounted on a styrofoam support in the open iris region common to both the input and the output waveguides. The bias field is applied along the common axis of the guides.

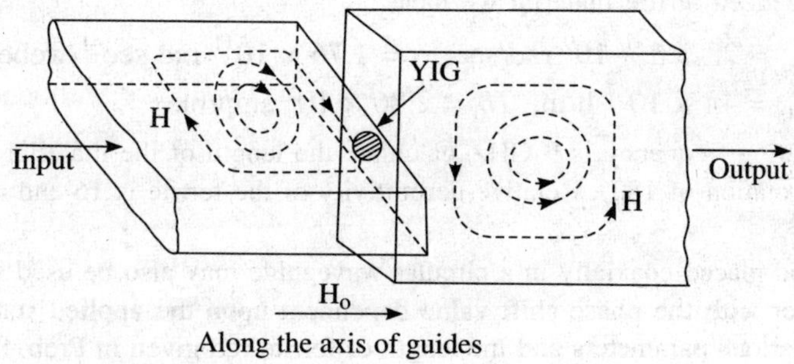

Fig. 8.18

YIG resonators find extensive use in tunable microwave filters and tunable microwave solid state oscillators.

SELECT BIBLIOGRAPHY

R. E. Collin, *Foundations for Microwave Engineering,* McGraw-Hill Book Company, 1966. Chapter 6 on Passive Microwave Devices.

B. Lax and K.J. Button, *Microwave Ferrites and Ferrimagnetics,* McGraw-Hill Book Company, New York, 1962.

R. F. Soohoo, *Theory and Applications of Ferrites,* Prentice-Hall Inc., Englewood Cliffs, N.J., 1960.

P. Clarricoats, *Microwave Ferrites,* John Wiley and Sons Inc., New York, 1961.

A. G. Gurevich, *Ferrites of Microwave Frequencies,* translated from Russian by A. Tybulewicz, Consultants Bureau, New York, 1963.

J. Roberts, *High Frequency Applications of Ferrites,* D. Van Nostrand Company Inc., Princeton, N.J., 1960.

P. S. Carter, 'Magnetically Tunable Microwave Filters Using Single Crystal YIG Resonators', *IRE Trans. MTT-9.* No. 3, May 1961, p. 252.

PROBLEMS

8.1 Write the scattering matrix for an ideal 4-port circulator.

8.2 (a) Can a 4-port circulator be obtained by combining two 3-port circulators?

(b) Can a 4-port circulator be used as a 3-port circulator?

8.3 Use relation (8.6) to study the second harmonic generation in a ferrite. Assume that $\mathbf{M} = \mathbf{M}_1\, ej\omega t + \mathbf{M}_2\, e^2 j\omega t$ for the AC magnetization and that $\mathbf{H}$ has only an x component with time dependence $ej\omega t$. Static magnetic field $\mathbf{H}_0$ is in z-direction. Neglect the third-harmonic term that occurs and show that

$$2j\omega\mathbf{M}_2 = \gamma\mu_0\, H_x\, [\mathbf{a}_x \times \mathbf{M}_1] + \gamma\mu_0\, H_0\, [\mathbf{a}_z \times \mathbf{M}_2]$$

where $\mathbf{a}_z$ represents a unit vector along z.

8.4 For a magnetized ferrite material we have :

$$\omega_m = 2\pi\, 5.6 \times 10^9 \text{ rad/sec}, \quad \gamma = 1.76 \times 10^{11} \text{ rad sec}^{-1}/\text{weber m}^{-2},$$

$$\mu_0 = 4\pi \times 10^{-7} \text{ h/m}, \quad H_0 = 2.86 \times 10^5 \text{ amp/m}.$$

If the operating frequency is 9 GHz, calculate the length of the material required for a Faraday rotation of 180°. Relative permittivity of the ferrite is 16 and $\varepsilon_0 = 8.854 \times 10^{-12}$ F/m.

8.5 A ferrite rod placed coaxially in a circular waveguide may also be used as a variable phase shifter with the phase shift value dependent upon the applied static magnetic field. For various parameters and the length of ferrite rod given in Prob. 8.4, calculate the change in the phase of the output signal when the static magnetic field changes by one per cent.

8.6 Using the expressions for field components in rectangular waveguide given in Chapter 2, derive equation (8.29).

8.7 Modify the waveguide-YIG filter configuration shown in Fig. 8.18, so that the off-band rejection is improved by the use of two YIG spheres. (The basic idea of the filter shown in Fig. 8.17 will be helpful.)

CHAPTER 9

Microwave Control and Logic Components

There are two types of control functions at microwave frequencies. These can be labelled 'analogue' and 'digital' respectively. In the 'analogue' family we include components like switches, phase shifters, attenuators, modulators, limiters, etc. These components are used very frequently in various microwave systems such as microwave communication links, radars, etc. Two main varieties of these 'analogue' control components are those using ferrites and those using PIN diodes. Principles of ferrite components have been discussed in the previous chapter. Ferrite switches, modulators and variable attenuators were mentioned. Operating principle of one of the possible types of ferrite phase shifter is brought out in Problem 8.5. In the present chapter we will confine ourselves to PIN diode circuits. The structure and the operation of PIN diodes is described in Sec. 9.1. Switches and phase shifters using PIN diodes are discussed in Sections 9.2 and 9.3, respectively. PIN diode attenuators, molulators and limiters are included in Sec. 9.4.

'Digital' control components at microwave frequencies are ultrafast logic circuits using narrow pulses with major part of their frequency spectrum in the microwave range. These components are useful in microwave instrumentation (e.g. frequency counters) and in data communication at gigabit rates. Sub-nanosecond pulses are also useful in several special purpose radars for docking, precollision sensing, etc. Two types of devices—transferred electron devices and GaAs MESFET's—are used for sub-nanosecond logic circuits. We will discuss the logic components briefly in Sections 9.5 and 9.6.

9.1 PIN DIODES

A PIN diode is a useful control element at microwave frequencies. These diodes differ considerably from ordinary *p-n* junctions. A PIN diode has a heavily doped *p*-region and a heavily doped *n*-region separated by a layer of high resistivity material that is nearly intrinsic. The thickness of the high resistivity layer is usually in the range of 10 to 200 μm. The doping profile for the device is shown in Fig. 9.1. The intrinsic layer has a resistivity of about 1000 ohm-cm and is either *p*-type or *n*-type. This lightly doped layer is called π-type or ν-type depending upon whether the conductivity is *p*-type or *n*-type, respectively.

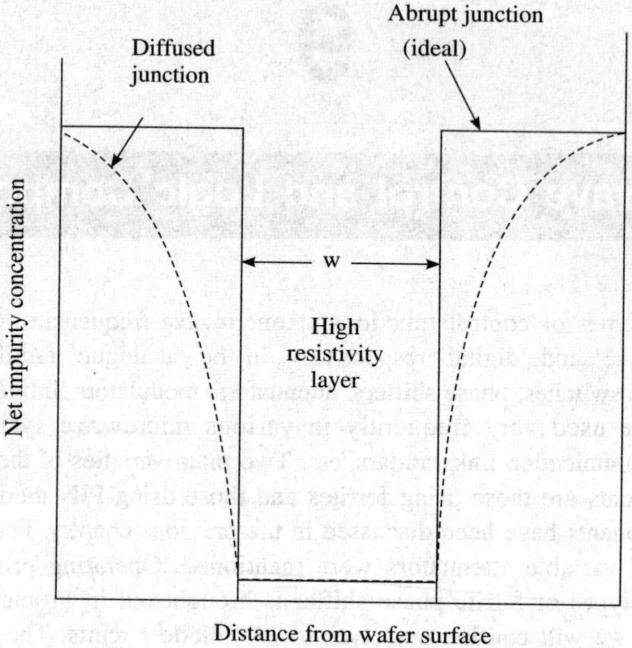

Fig. 9.1

At zero bias, two space-charge regions are formed in the *p* and *n* layers adjacent to the intrinsic layer because of the diffusion of holes and electrons across the junctions. The thickness of these regions is inversely proportional to the impurity concentration. In the case of an ideal diode, the intrinsic layer has no impurities, i.e. it is totally depleted of mobile charge carriers. Thus, we have a region of fixed negative charge in the *p*-layer and a region of fixed positive charge in the *n*-layer with equal charges in the two regions. When a reverse bias is applied, the space-charge regions in the *p* and *n* layers become wider. A uniform electric field exists in the intrinsic region, dropping linearly to zero through the depletion regions in the *n* and *p* layers. This situation is represented in Fig. 9.2.

When the high-resistivity layer is actually a π-layer, the situation is somewhat different and is depicted by dotted lines in Fig. 9.2. With no applied bias, the diffusion of holes and electrons across the *n*-π junction produces a very thin depletion region in the *n*-layer and a thicker depletion region in the π-layer. The two depletion regions contain equal but opposite fixed charges. In this case, the π-layer is not completely depleted of mobile carriers. As the reverse bias is applied, the depletion region in π-layer becomes more and more thick till the entire π-layer is swept free of mobile carriers. The small negative voltage required for this purpose is called the bias required to sweep out the π-region. With increasing reverse bias, a thin depletion region appears in *p*-layer also. The electric field rises steeply in the *n*-region, decays slowly across the π-region and then drops steeply to zero in the *p*-depletion region.

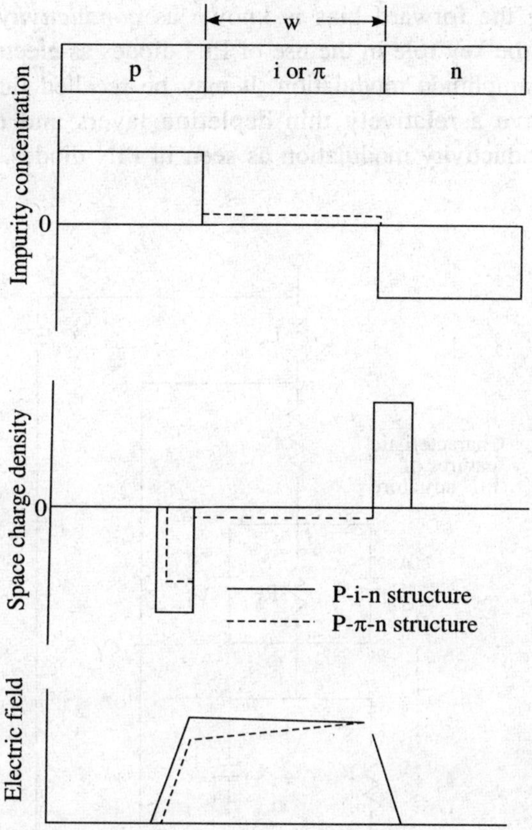

Fig. 9.2

When a forward bias is applied to the diode, carrier injection into the π-layer takes place. Electrons are injected into the π-layer from the *n*-layer, and holes are injected from the *p*-layer. The carriers diffuse into the π-layer, their concentration diminishing with depth into the π-layer because of recombination. The diffusion of carriers causes the carrier concentrations in the π-layer to increase above their equilibrium levels, and the resistivity of this layer drops as the forward bias is increased. The hole and electron concentrations are about equal throughout the π-layer. If the carrier lifetimes in the π-layer are relatively long and the π-layer is not too thick, the π-layer becomes flooded with carriers at a reasonable forward bias level. When this happens, the diode exhibits a low resistance and, in fact, appears as a virtual short circuit across a microwave transmission line. For this characteristic of the PIN diode, it is essential that the lifetime of carriers in the π-layer be greater than the time period at the operating frequency. For this reason PIN diodes cannot be operated usefully at lower radio frequencies.

The effect of controlling the carrier density, and hence the resistance in the i-layer, by varying the forward bias is known as ***conductivity modulation.*** This phenomenon plays the key role in the use of PIN diodes as electronically controlled attenuators and in amplitude modulation. It may be recalled that the ordinary p-n junction diodes have a relatively thin depletion layers and do not exhibit the phenomenon of conductivity modulation as seen in PIN diodes.

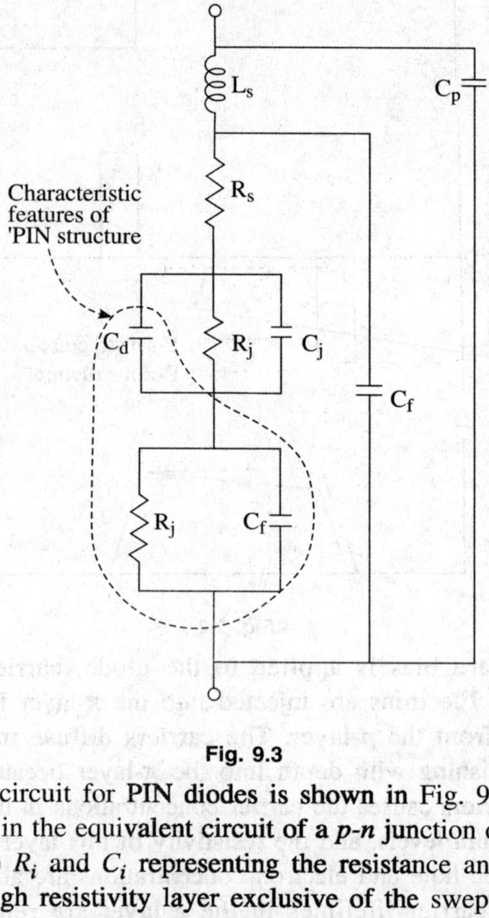

Fig. 9.3

An equivalent circuit for PIN diodes is shown in Fig. 9.3. In addition to all the elements present in the equivalent circuit of a ***p-n*** junction diode, this equivalent circuit contains : (a) R_i and C_i representing the resistance and the capacitance of the portion of the high resistivity layer exclusive of the swept out region, and (b) diffusion capacitance C_d representing the charge storage caused by the current flow through the intrinsic region. These characteristic features of a PIN structure are shown enclosed in a dotted loop in Fig. 9.3. Other components in the equivalent circuit correspond to the model of a p-n junction diode. R_j and C_j are junction resistance and junction capacitance, respectively. R_s is the ohmic series resistance of the diode and C_f is the fringing capacitance associated with the device chip. Series inductance of the bonding wire is represented by L_s, and the capacitance C_p accounts for the package capacitance.

Under a small reverse bias, when only a part of the high resistivity layer is swept out, the equivalent circuit may be simplified by taking R_j much greater than X_j (= $1/\omega C_j$) and R_i greater than X_i (= $1/\omega C_i$). When the reverse bias is enough to sweep out the high resistivity layer completely, we have both X_i and R_i equal to zero. A simplified equivalent circuit under these conditions is shown in Fig. 9.4(a). The diffusion capacitance C_d is zero under reverse bias condition.

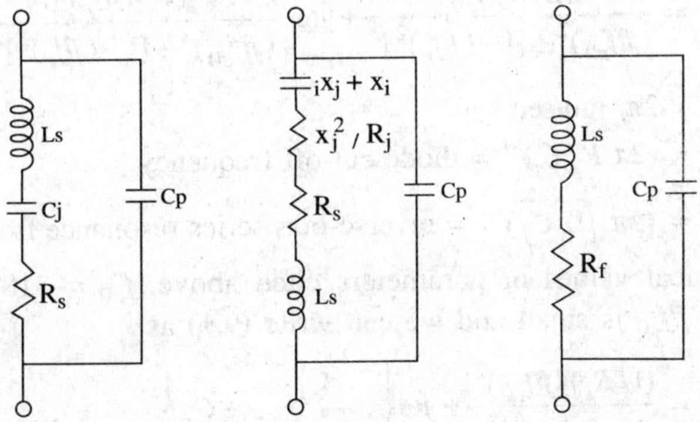

Fig. 9.4

At zero bias, R_j can be ignored and the parallel combination of R_j and C_j can be represented by a resistance X_i^2/R_i, in series with the diode capacitance as shown in Fig. 9.4 (b).

Under forward bias condition, the diode resistance R_f consists of the intrinsic layer resistance R_i and the series resistance R_s. R_i decreases with the forward bias. The equivalent circuit can now be expressed as shown in Fig. 9.4 (c).

These equivalent circuit representations are useful in the analysis and design of PIN diode circuits.

9.2 PIN DIODE SWITCHES

Performance of switches may be represented in terms of two important parameters, insertion loss and isolation. Insertion loss is the ratio (in dBs) of the power incident at the input to the power transmitted by the switch when in 'on' condition. Isolation is a measure of power transmitted by the switch when it is in 'off' condition. It is defined as the ratio (in dBs) of the power incident at the input to the power transmitted with the switch in 'off' condition.

As pointed out in Sec. 9.1, PIN diode has a very high resistance when in reverse bias and a very low resistance when in the forward bias. Typical values are greater than 10 kilo-ohms and less than one ohm, respectively. Since these values are much higher or much lower than commonly used transmission line impedances, the PIN diode behaves as an open-circuited and a short-circuited element under two bias conditions, respectively.

9.2.1 Diode parameters

The diode equivalent circuits employed in subsequent analysis are shown in Figs. 9.4 (a) and (c) for the reverse and the forward bias conditions, respectively. Typical values of various elements are: R_s = 1.0 ohm, C_j = 0.5 pF, R_f = 1 ohm, L_s = 0.5 nH and C_p = 0.3 pF. Using the reverse bias equivalent circuit, we obtain the following expression for the reverse bias admittance

$$Y_r = \frac{(1/R_s)(f/f_{c0})^2}{(f/f_{c0})^2+[1-(f/f_r)^2]^2} + j\omega\left\{\frac{C_j[1-(f/f_r)^2]}{(f/f_{c0})^2+[1-(f/f_r)^2]^2}+C_p\right\} \quad ...(9.1)$$

where $\omega = 2\pi f$ rad/sec

$f_{c0} = (2\pi\, R_s\, C_j)^{-1}$ = diode cut-off frequency

and $f_r = (2\pi\sqrt{L_s\, C_j})^{-1}$ = reverse-bias series resonance frequency

For the typical values of parameters cited above, f_{c0} = 318 GHz and f_r = 10 GHz. Usually f/f_{c0} is small and we can write (9.1) as

$$Y_r = \frac{(1/R_s)(f/f_{c0})^2}{[1-(f/f_r)^2]^2} + j\omega\left\{\frac{C_j}{1-(f/f_r)^2}+C_p\right\} \quad ...(9.2)$$

Also, diodes are commonly selected such that $(f/f_r)^2 \ll 1$, and (9.2) may be written as

$$Y_r = G + jB = (1/R_s)(f/f_{c0})^2 + j\omega\,[C_j + C_p] \quad ...(9.3)$$

Expression for the impedance under the forward bias (Fig.9.4c) can be put in a simple form when $(\omega C_p) \ll (\omega L_s)^{-1}$ and we have

$$Z_f = R + jX = R_f + j\omega L_s \quad ...(9.4)$$

For switching applications PIN diodes may be mounted across the transmission lines in two possible configurations. These are called shunt-mounted and series-mounted configurations and are shown in Fig. 9.5 (a) and (b), respectively. Biasing circuits for the diodes are not shown in these figures. These two types of switches are discussed in the following way:

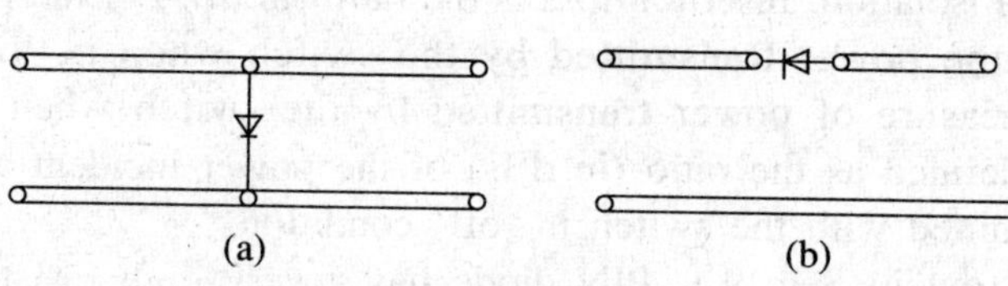

Fig. 9.5

9.2.2 Shunt mounted switch

A shunt mounted switch is 'on' when the diode is reverse biased (high impedance) and 'off' when the diode is forward biased (low impedance). These two cases may be analysed by using simplified equivalent circuits shown in Figures 9.6 (a) and (b), respectively. Values of $(G + jB)$ and $(R + jX)$ are given by equations (9.3) and (9.4), respectively.

For calculation of insertion loss, one can consider the circuit of Fig. 9.7 (a). This corresponds to the case when the source and the load admittances are equal to the characteristic admittance of the transmission line. For an ideal switch, the power delivered to the load is given by

$$P_0 = I^2/Y_0 \qquad ...(9.5)$$

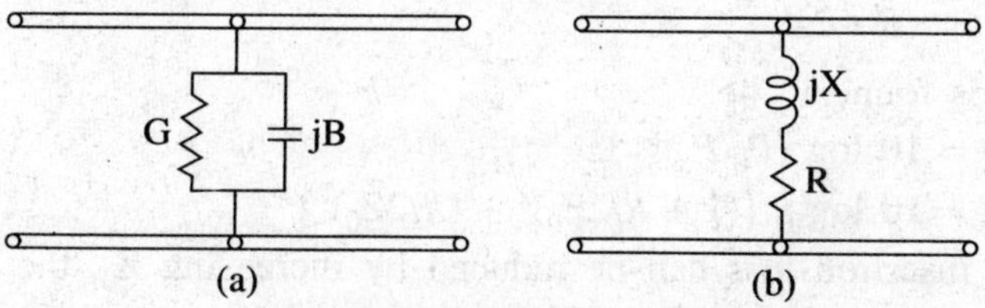

Fig. 9.6

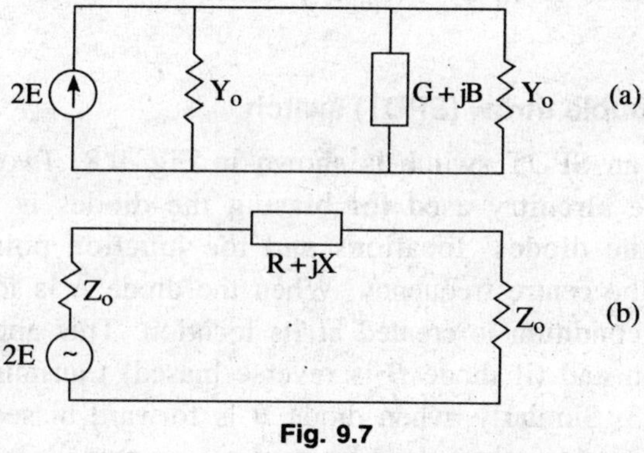

Fig. 9.7

Power delivered to the load, when the diode admittance $(G + jB)$ is taken into account, is given by

$$P_d = (2IY_0)^2/\{Y_0[(2Y_0 + G)^2 + B^2]\} \qquad ...(9.6)$$

From equations (9.5) and (9.6) insertion loss I_L may be written as

$$I_L = 10\log_{10}\{(1 + G/2Y_0)^2 + (B/2Y_0)^2\} \qquad ...(9.7)$$

This relation indicates that the insertion loss I_L may be reduced by increasing Y_0, i.e. by reducing the characteristic impedance of the transmission line. Also, the insertion loss may be reduced by decreasing B. The susceptance B can be made zero by adding an inductive susceptance in parallel such that it forms a resonant circuit at operating frequency.

Equation (9.7) gives the value of the isolation when G and B corresponding to the forward bias equivalent circuit (Fig. 9.4c) are used.

9.2.3 Series mounted switch

This switch configuration is shown in Fig. 9.6 (b). The switch is 'on' when the diode is forward biased and it is 'off' when the diode is reverse biased. Insertion loss I_L may be calculated by considering the circuit shown in Fig. 9.7(b). Power transferred to the load for an ideal switch case is given by

$$P_0 = E^2/Z_0 \qquad ...(9.8)$$

when the finite value of the diode impedance is considered, the power transmitted to load becomes

$$P_d = \frac{4E^2 Z_0}{(R+2Z_0)^2 + X} \qquad ...(9.9)$$

Insertion loss I_L is found to be

$$I_L = 10 \log (P_0/P_d)$$
$$= 10 \log_{10} \{(1 + R/2Z_0)^2 + (X/2Z_0)^2\} \qquad ...(9.10)$$

In this case, the insertion loss can be reduced by increasing Z_0, the characteristic impedance of the transmission line. Also, the insertion loss may be reduced by reducing the reactance X by adding an external capacitor in series with the diode. This capacitance value is given by $1/(\omega_0^2 L_s)$, where ω_0 is the operating frequency (in radians).

9.2.4 Single-pole double throw (SPDT) switch

A configuration for an SPDT switch is shown in Fig. 9.8. Two PIN diodes are used. As before, the circuitry used for biasing the diodes is not shown. The distances between the diodes' locations and the junction point J are quarter wavelength each at the centre frequency. When the diode A is forward biased an almost short circuit condition is created at its location. This appears as an open circuit at the junction and (if diode B is reverse biased) transmission takes place from port 1 to port 3. Similarly, when diode B is forward biased and diode A is reverse biased, transmission takes place from port 1 to port 2.

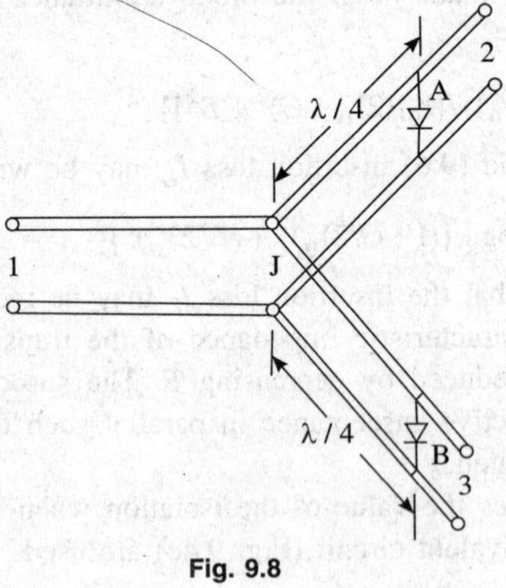

Fig. 9.8

It may be noted that the short circuited branch with the forward biased diode appears as a quarter wave stub as shown in Fig. 9.9. This feature causes the switch characteristics to vary with frequency even when the diodes are ideal. It may be recalled that the SPST switches discussed earlier have frequency limitation only because of the diode reactances.

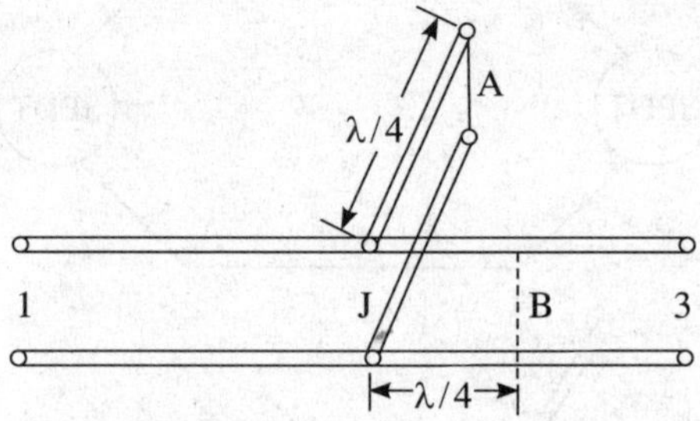

Fig. 9.9

Based on the principle of SPDT switch discussed above, single-pole multiple throw (SPMT) switches may also be designed. An SPNT switch needs N diodes and all the $(N + 1)$ lines arc connected together at the junction J. For an accurate analysis, discontinuity reactances at the junction should be taken into account.

9.3 PHASE SHIFTERS

Phase shifter is a two-port network in which phase shift of a signal travelling from one port to the other can be controlled. Electrically controlled phase shifters use either diodes or magnetized ferrites. Only diode phase shifters are discussed in this section.

Two types of phase shifters are useful. When the phase shift can be changed continuously from one value to another, the device is called a continuously variable (or analogue) phase shifter. This type of phase shifters are used in microwave bridges and instruments. In the other type of phase shifters only discrete values of the phase shift are available. These are called digital phase shifters and are used extensively in phased array antennas.

Digital phase shifters may be designed by using switches and transmission line sections. Two possible configurations, transmission type and reflection type are shown in Figs. 9.10 (a) and 9.10 (b and c), respectively. Transmission type phase shifter uses two SPDT switches. Depending upon the bias applied to the two switches, the signal transmission can follow either path A or path B. Path B is longer than path A by a distance l. When the signal is switched from the path A to the B, the output undergoes a phase shift of βl radians, where β is the phase constant along the line. Thus two discrete values of phase shift are possible. This is called a one-bit digital phase shifter.

Reflection type digital phase shifter shown in Fig. 9.10 (b) uses a circulator and an SPST switch. Alternatively, a 3 dB directional coupler and two SPST switches may be employed as shown in Fig. 9.10 (c). In Fig. 9.10 (b), when the switch is open an input at port 1 gets coupled to port 2, travels a distance $B/2$ to

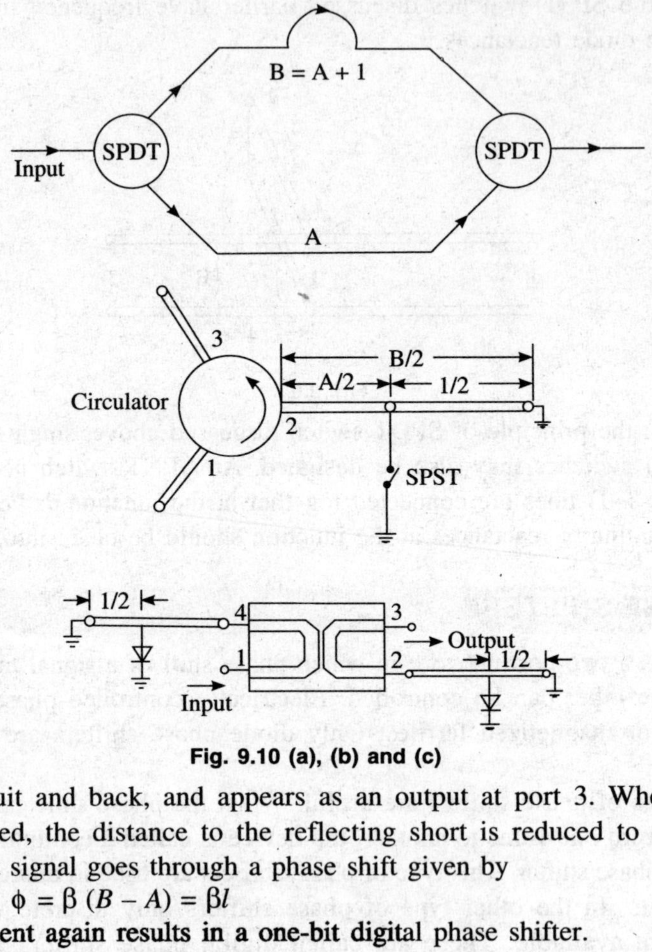

Fig. 9.10 (a), (b) and (c)

the short circuit and back, and appears as an output at port 3. When the SPST switch is closed, the distance to the reflecting short is reduced to a length $A/2$ and the output signal goes through a phase shift given by

$$\phi = \beta (B - A) = \beta l \quad ...(9.11)$$

This arrangement again results in a one-bit digital phase shifter.

A continuously variable phase shifter may be designed by terminating a line in a variable reactance as shown in Fig. 9.11.

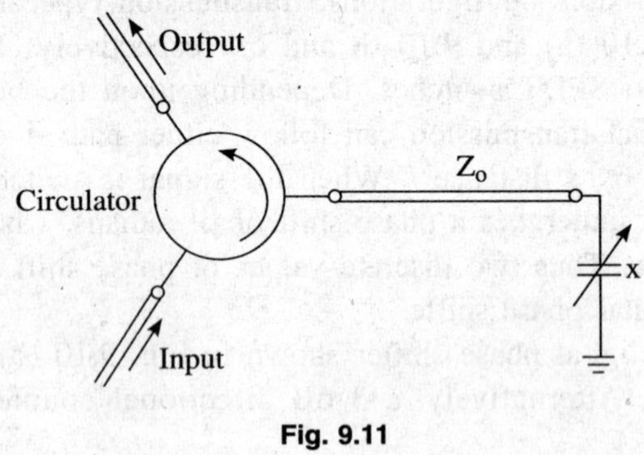

Fig. 9.11

Reflection coefficient when a transmission line of characteristic impedance Z_0 is terminated in a reactance X is given by

$$\rho = \frac{jX - Z_0}{jX + Z_0} \quad ...(9.12)$$

Reflection phase change is given by

$$\phi = 2\tan^{-1}\frac{-X}{Z_0} \quad ...(9.13)$$

If a varactor diode is used the value of X can be varied from X_1 to X_2 by varying the reverse bias applied to the diode. This will change the phase shift continuously from ϕ_1 to ϕ_2. Values of ϕ_1 and ϕ_2 are calculated by using the relation (9.13) and the characteristics of the varactor diode.

9.4 PIN ATTENUATORS, MODULATORS AND LIMITERS

9.4.1 Attenuators

As discussed in Sec. 9.1, the resistance of the intrinsic layer of a forward biased PIN diode is a strong function of the bias current. This property is used in designing current controlled variable attenuators at microwave frequencies. Like other PIN diode circuits the lowest frequency of operation of these attenuators is determined by the minority carrier lifetime and for lifetimes around 100 ns it is typically 10 MHz.

It is found that under the forward bias condition the intrinsic layer resistance of a typical PIN diode varies according to the following relationship*

$$R_i = 26\ I^{-0.87} \quad ...(9.14)$$

where I is the forward bias current in mA. This relationship is good for the range of bias current between 19 μA and 10 mA. Above 10 mA, the forward resistance R_F approaches parasitic series resistance R_S (typically 0.5 to 0.1 ohm). Towards zero bias and reverse bias, a limit of typically 10,000 ohms is asymptotically approached.

An important characteristic of a variable attenuator is that its input impedance should remain constant so that the attenuator stays matched over its operating range.

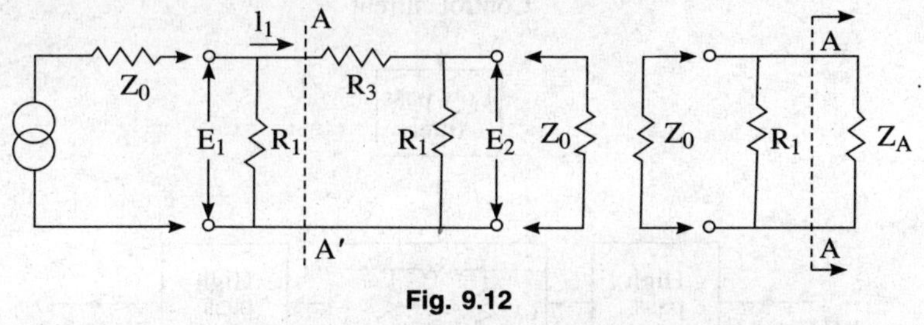

Fig. 9.12

* Value of the constant 26 in eqn. (9.14) depends upon the width of *i*-layer average mobility and lifetime of carriers. See H.P. Application Note No. 932 listed in Select Bibliography.

One way of realizing this is to have a pi-network shown in Fig. 9.12 and consisting of three PIN diodes for R_1, R_3, R_1, respectively. Let the impedance of the network to the right *of AA'* be indicated as Z_A, where $Z_A = (R_3 + R_1 Z_0/\{R_1 + Z_0\})$. For the input impedance matching, the input impedance of the network must equal the source impedance. Therefore,

$$\frac{1}{Z_0} = \frac{1}{R_1} + \frac{1}{Z_A} \qquad ...(9.15)$$

We have $E_1 = I_1 Z_A$ and $E_2 = I_1 R_1 Z_0/(R_1 + Z_0)$, so the attenuation ratio E_1/E_2 may be written as

$$\frac{E_1}{E_2} = \frac{Z_A}{R_1 Z_0/(R_1 + Z_0)} \qquad ...(9.16)$$

Substituting for Z_A from (9.15) we have

$$K = E_1/E_2 = (R_1 + Z_0)(R_1 - Z_0) \qquad ...(9.17)$$

R_1 and R_3 can be expressed in terms of the characteristic impedance of the network (Z_0) and the input to the output voltage ratio K as follows :

$$R_1 = Z_0 (K + 1)/(K - 1) \qquad ...(9.18a)$$

and

$$R_3 = Z_0 (K - 1/K)/2 \qquad ...(9.18b)$$

Equations (9.18) give the variation of pi-arm resistances with respect to input-output voltage ratio *K*. Using (9.14) we can find the bias currents needed to produce the necessary resistance values for R_1 and R_3.

The complete design of the attenuator includes, in addition to the above, mounting the PIN diodes in an RF transmission line structure and the necessary DC biasing circuitry for the control of the PIN diode resistances.

9.4.2 PIN modulators

When an absorption-type PIN attenuator discussed above is placed along a transmission line, sine-wave, square-wave, and pulse modulation can be obtained by varying currents through the diodes in a suitable manner. As the input impedance

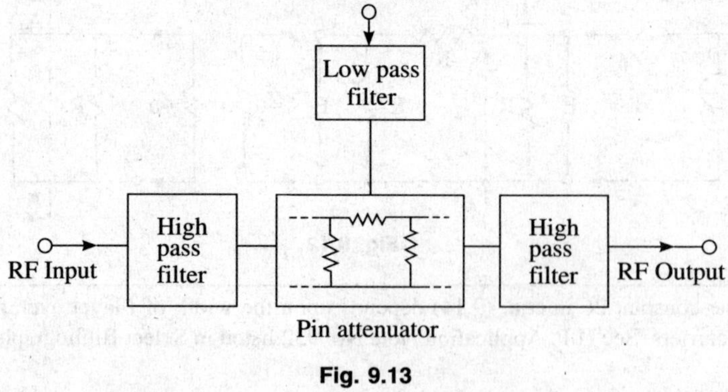

Fig. 9.13

of the attenuator does not change with the value of attenuation, frequency or the output of the microwave source (at the input) does not get disturbed by the modulator. The PIN modulators are three-port devices. One of the ports is for the modulating signal and the other two are RF ports. Unmodulated input can be connected to any one of the two RF ports. A typical block diagram is shown in Fig. 9.13. A typical application of the PIN modulator is found in closed-loop automatic levelling circuits (ALC) discussed in Sec. 6.4.

9.4.3 PIN limiters

Limiters are used for the protection of microwave systems containing sensitive amplifiers, mixers, detectors, samplers, etc., when these circuits are operated in environments which may expose them to high levels of microwave power. The two basic types of limiters are feedback systems and passive limiters. A feedback limiter system is an ALC loop. In a passive limiter the same element detects the RF power and causes limiting. The three types of passive limiters commonly used are gas tubes, ferrite limiters and diode limiters. We will discuss PIN diode limiters only. These limiters can typically handle the power levels of about 100 watts peak and react in a few nanoseconds.

A PIN diode limiter is a microwave switch that is controlled by self bias rather than an externally applied bias. PIN diodes with very short carrier lifetimes (typically less than 1 ns) are used to ensure that they will efficiently rectify the RF power when it exceeds a specified level. Limiters use shunt-mounted diodes, and the presence of excess power (and therefore bias) switches off the power. A typical limiter circuit is shown in Fig. 9.14 (a). Two PIN diodes are connected in shunt across a transmission line. The diodes are spaced $\lambda/4$ apart so that the individual reflections cancel and a low VSWR is obtained. The choke shown provides a return path for the self bias current. A curve of limiting characteristics is shown in Fig. 9.14 (b).

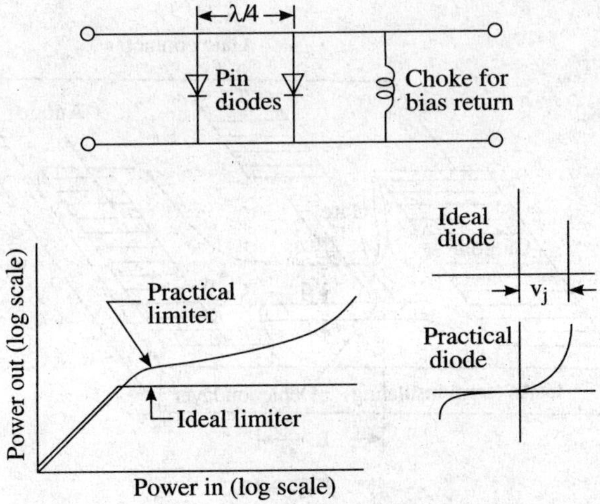

Fig. 9.14 (a) and (b)

At limiting, threshold voltage across the diode is equal to built-in junction potential V_j. At high levels, voltage across the diode exceeds the built-in junction potential V_j and diode resistance is very low. The circuit acts as a closed switch and incident power is reflected back. The input-output characteristic of a practical limiter deviates from the ideal curve because the transition from high reverse to low forward resistance is not abrupt.

9.5 LOGIC CIRCUITS USING TRANSFERRED ELECTRON DEVICES

Digital logic operations at microwave frequencies may be carried out by using suitably designed *transferred electron devices* or by GaAs MESFET devices. Logic circuits using TEDs are discussed in this section and those using MESFETs are described in Sec. 9.6.

9.5.1 Gunn effect digital devices

As discussed in Chapter 4, transferred electron devices exhibit two distinct states, one for the bias voltages below threshold when the device behaves like a resistor and the other for the bias above threshold when high field domains are present. Also, the threshold field is higher than the field required to sustain the existence of high field domains. These characteristics make TEDs suitable for logic circuits. Various techniques of designing these circuits have been described in literature. However, the most promising of these appears to be the use of Schottky-barrier-gate Gunn-effect-digital devices (SBG-GEDD).

SBG-GEDD is a planar device with a configuration shown in Fig. 9.15. An *n*-type GaAs active layer is epitaxially grown on a semi-insulating substrate. Cathode and anode contacts are made broader than the active region. A Schottky barrier (SB) gate is placed near the cathode. When this SB gate is reverse biased, the depletion layer stretches to concentrate the current path under the gate.

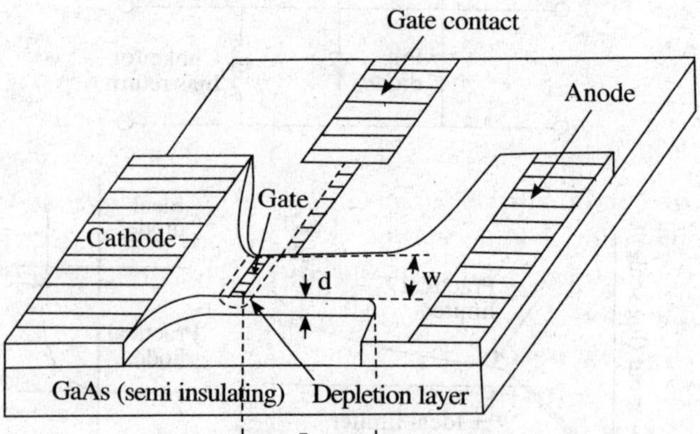

Fig. 9.15

This causes the electric field to exceed the threshold level. A high field domain is formed under the gate edge near the cathode and propagates towards the anode. When the bias is just below the threshold field, a domain can be triggered by a small input pulse at the gate.

Voltage-current relationship for the device is shown in Fig. 9.16. The thereshold voltage V_{th} and the threshold current I_{th} decrease with gate voltage, while the valley current I_v is constant. In most of the logic circuits the device is biased just below the threshold at zero gate voltage. Therefore, bias current I_b is nearly equal to threshold current I_{th}. The device is stable at a point Q. Then by a small negative input voltage, the stable point moves from Q to Q' at the intersection of the load line and $I = I_v$. The point Q' is stable only during a domain transit. Extinction of the domain at the anode makes the operating point to move to Q again. Thus by a small input pulse applied to the gate, a wavy current pulse is obtained at the anode (and also at the cathode). The amplitude of the current pulse is approximately $(I_{th} - I_v)$ and the duration is L/v, where L is the distance from the gate to the anode and v is the domain velocity ($\simeq 10^7$ cm/sec). The rise and fall times of the pulse are determined by the formation and extinction times of domains and are of the order of 10 picoseconds. Thus SBF-GEDDs are capable of generating and amplifying sub-nanosecond pulses. Also, these devices can be used for combinatorial and sequential logic circuits.

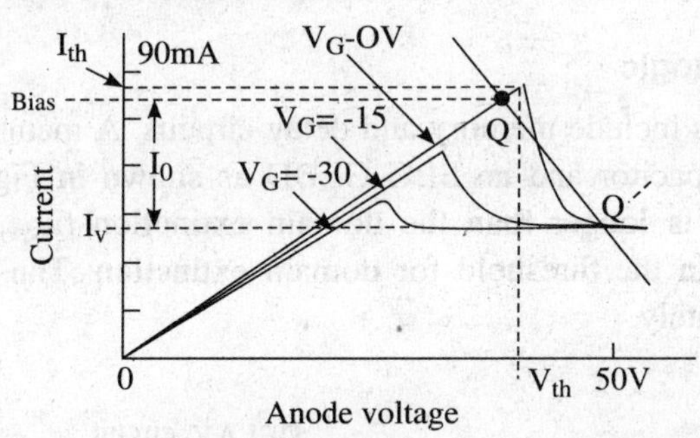

Fig. 9.16

9.5.2 Combinatorial logic

Two examples of combinatorial logic circuits using SBG-GEDD are shown in Fig. 9.17. In combinatorial logic circuits a 'weighting SUM' is a basic function. Fig. 9.17(a) shows a circuit suitable for this. When inputs at X_1 and X_2 are separated by a time interval longer than the output pulse duration, two pulses appear at the output. When two inputs are present at the same time, the amplitude of the output pulse becomes twice. For 'weighting', the current through each device is varied according to the weighting factor by changing the device area appropriately.

Another important logic function is 'INHIBIT' or 'NOT'. A circuit for this function is shown in Fig. 9.17 (b). Let us consider a negative pulse as signal '1' and a positive pulse or zero level as signal '0'. A negative pulse at *U* shifts the operating point of SBG-GEDD 1 to *Q'* with $I \simeq I_v$, and results in a negative pulse at the output *X*. When a negative pulse is applied to L, SBG-GEDD 2 operates at *Q'* and a positive output ('0' signal) appears at *X*. Under this condition the voltage across SBG-GEDD 1 is much below threshold and a signal at *U* is not able to trigger it. Thus the signal at *U* is inhibited from appearing at the output.

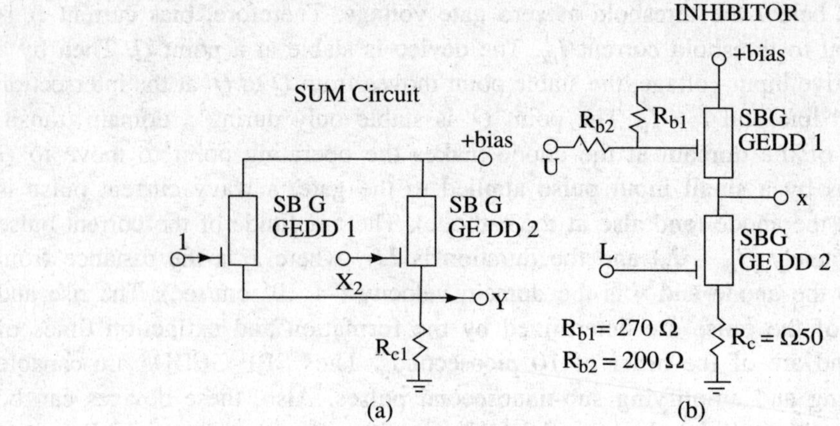

Fig. 9.17

When there is no input at *U*, the circuit works as 'NOT' for input at *L*.

9.5.3 Sequential logic

These logic circuits include memory and delay circuits. A memory circuit is realized by a resistor, a capacitor and an SBG-GEDD as shown in Fig. 9.18(a). When the RC time constant is longer than the domain extinction time, field in the device remains higher than the threshold for domain extinction. Thus the next domain is nucleated immediately.

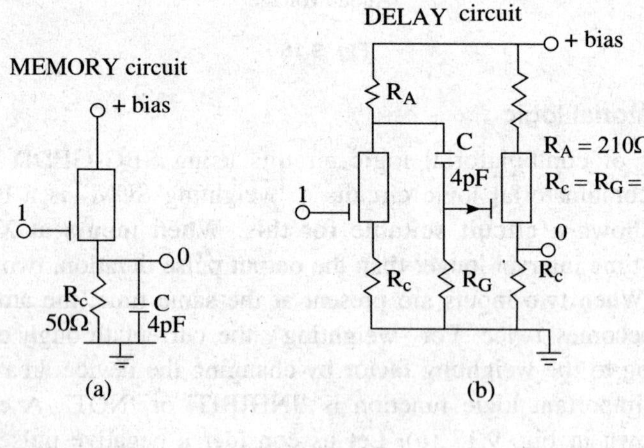

Fig. 9.18

Therefore, the device oscillates when a negative pulse is applied to the gate. When a positive pulse is applied to the gate, the oscillations stop. Thus, negative and positive pulses act as 'set' and 'reset' signals, respectively.

A delay circuit is shown in Fig 9.18(b). The second SBG-GEDD is triggered by time-differential of the trailing edge of a positive pulse at the anode first SBG-GEDD. The delay time is equal to the width of an output pulse from the first SBG-GEDD. Combination of memory and delay circuits can be used for designing larger sequential circuits like shift registers, etc.

9.6 LOGIC CIRCUITS USING GaAs MESFETs

The principle of operation of MESFETs has been discussed in Sec. 4.2. GaAs MESFETs are capable of operating at high frequencies which are at present beyond the range of bipolar transistors. Also, the GaAs MESFET is a potentially useful three-terminal switching device with very fast response and can be monolithically integrated along with resistors and Schottky diodes to form high speed logic circuits. Monolithic integrated-logic gates with 100 ps signal propagation delay (fan out 2) and 4 pJ speed-power product have been built. A digital frequency divider operating at 4 GHz has been demonstrated. Also, a pulse regenerator and amplifier for pulses with rise time in 50 ps range has been reported.

A typical set-up for a simple MESFET logic is shown in Fig. 9.19. In this design, the positive NOR function is realized by parallel combination of MESFETs, and the NAND function by series combination in the form of a dual-gate device. In a dual-gate MESFET, a negative bias on any one of the gates causes the device to be cut off.

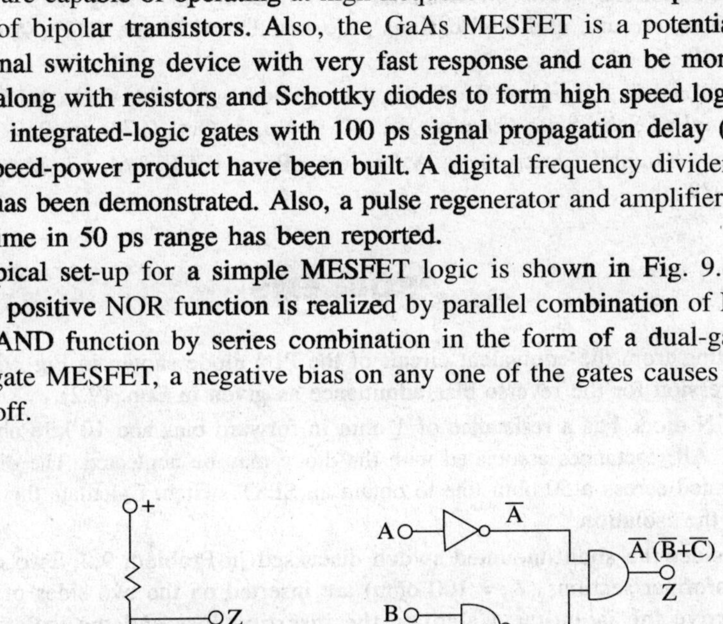

Fig. 9.19

It has been pointed out that by combining SBG-GEDDs with MESFETs extremely fast digital logic circuits can be designed. These logic circuits are useful in microwave instrumentation. For example, digital frequency division at 4 GHz can be employed for a microwave frequency counter. Also, sub-nanosecond pulses, which can be formed and processed using these logic circuits, are useful in several special purpose radars.

SELECT BIBLIOGRAPHY

H.A. Watson, *Microwave Semiconductor Devices and their Circuit Applications,* McGraw-Hill Inc., 1969. Chapter 9 on 'PIN Diodes' and Chapter 10 on 'Microwave switches, limiters and phase shifters'.

K.C. Gupta and A. Singh, *Microwave Integrated Circuits,* Wiley Eastern, New Delhi, 1974. Section 8.6 on 'PIN diodes' and Chapter 10 on 'Design of microstrip circuits'.

Hewlett Packard, 'Selection and use of microwave diode switches and limiters', Application Note 932, May 1973.

——, 'An attenuator design using PIN diodes', Application Note 912.

——, 'The PIN diode as a microwave modulator', Application Note 58, Nov. 1968.

T. Sugeta *et al.*, 'Gunn-effect digital devices for sub-nanosecond pulse generators and logic functions', Proc. IV Biennial Cornell Conf. on 'Microwave semiconductor devices, circuits and applications', Cornell Univ. Ithaca (NY), Aug. 1973, pp. 201-10.

R.L.V. Tupl and C.A. Liechti, 'High speed integrated logic with GaAs MESFETs', *IEEE Journal of Solid-State Circuits,* Vol. SC-9, No. 5, Oct. 1974, pp. 269-76.

B.G. Bosch, 'Gigabit Electronics—A Review', Proc. IEEE, Vol. 67, March 1979, pp. 340-79.

PROBLEMS

9.1 Starting from the equivalent circuit of the PIN diode shown in Fig. 9.4, derive the expression for the reverse bias admittance as given in Eqn. (9.2).

9.2 A PIN diode has a resistance of 1 ohm in forward bias and 10 kilo-ohm in reverse bias. All reactances associated with the diode may be neglected. The diode is shunt-mounted across a 50 ohm line to obtain an SPST switch. Calculate the insertion loss and the isolation.

9.3 Consider the shunt-mounted switch discussed in Problem 9.2. Two quarter wave transformer sections (Z_0 = 100 ohm) are inserted on the two sides of the diode to improve the isolation. Calculate the insertion loss and the isolation with the transformers in the circuit.

9.4 A shunt-mounted PIN diode SPST switch is combined with a series-mounted switch. The two PIN diodes used are identical to the diode in Prob. 9.2. Find out the transmission line impedance that will lead to the minimum attenuation of the combined switch. What is the value of corresponding attenuation?

9.5 PIN diode of Problem 9.2 is used in an SPDT switch of the type shown in Fig. 9.8. Again, lines with 50 ohms characteristic impedance are used. Calculate the input VSWR at centre frequency when ports 2 and 3 are terminated in matched loads.

9.6 Consider the SPDT switch considered in Prob. 9.5. The operating frequency is increased by 5 per cent. Calculate the input VSWR again. A Smith chart may be used for calculations.

9.7 A reflection type phase shifter shown in Fig. 9.10 (b) is designed at 3 GHz. The diode used can be represented by a series combination of 1 ohm resistor and a 1 nH inductor while in forward bias, and by a parallel combination of 10 kilo-ohms resistor and 1 pF capacitor when in reverse bias. Calculate the differential phase shift that will be obtained if the distance (*B–A*) is exactly quarter-wavelength at the operating frequency. Use Smith chart. Discuss why the relation (9.11) is not valid in this case.

9.8 A 3 dB attenuator is to be designed for a 50 ohm system. Calculate the values of the resistors in the pi-network needed for this purpose. If values of R_1 and R_3 can vary from 2 ohms to 1 kilo-ohm, what is the range of attenuation that can be obtained from the matched attenuator?

9.9 A PIN diode can be represented by a series combination of 1 ohm resistor and a 1 nH inductor while in forward bias, and by a parallel combination of 10 kilo-ohm and 1 pF capacitor when in reverse bias. This diode is used for a series-mounted SPST switch using a 50 ohm transmission line. Calculate the insertion loss and isolation at 1 GHz and at 4 GHz.

9.10 The PIN diode described in Prob. 9.9 is now used in a shunt-mounted SPST configuration in a 50 ohm line. Calculate the values of insertion loss and isolation in this case.

9.11 An SPDT switch can also be designed using series-mounted diodes as shown in Fig. 9.20. Compare the VSWR performance of this circuit with the SPDT switch in Prob. 9.5 when identical diodes and line impedances are used.

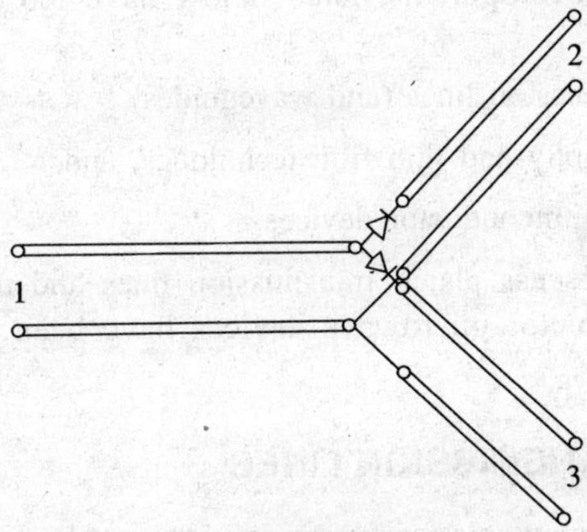

Fig. 9.20

9.12 A reflection type phase shifter having the configuration shown in Fig. 9.10 (c) uses a 3 dB coupled line directional coupler discussed in Sec. 7.3.3. Coupler characteristics are described by Eqns.(7.78) and (7.102). Find the relationship between output and input when diodes are (a) on, and (b) off. Diodes may be assumed ideal.

9.13 A varactor, whose capacitance can be varied from 1 pF to 0.5 pF by varying the DC bias, is used to design a continuously variable phase shifter as shown in Fig. 9.11. The characteristic impedance Z_0 is 50 ohms. Calculate the amount of phase shift that can be obtained at a frequency of 10 GHz.

CHAPTER 10

Microwave Integrated Circuits

Microwave Integrated Circuits (MIC) represent an extension of integrated circuit technology to extremely high frequencies (uhf and higher). An MIC is a subsystem function, which can vary greatly in complexity, and usually represents a major part of the microwave assembly of a specific system.

Today's MICs are mainly of a hybrid nature, in which active devices (and some passive devices) in chip form are mounted on substrates on which the remaining circuit elements have been fabricated using thin films. Monolithic MICs, although feasible presently for a few specific uses, represent a future generation of the MIC approach. We will discuss only hybrid MICs.

Three types of developments listed below have led to the present MIC technology:

(i) planar transmission lines (and waveguides),

(ii) photolithography and thin film techniques, and

(iii) microwave semiconductor devices.

We will briefly discuss planar transmission lines and thin film technology. Some of the semiconductor microwave devices have been discussed earlier in Chapters 4 and 9.

10.1 PLANAR TRANSMISSION LINES

Planar transmission lines suitable for microwave integrated circuits have a geometry that allows the characteristic impedance of the line to be controlled by defining the dimensions in a single plane. This feature allows the complete transmission line circuit to be fabricated in one step by thin film technology and photolithography techniques. These processes are similar to those used for making printed circuit cards for low frequency electronic circuitry.

Several configurations for planar transmission lines are possible. Commonly used are stripline, microstrip, slotline and coplanar lines. Many of these are not TEM-mode lines. Some principal features of these lines are described below.

10.1.1 Stripline

Stripline is a three-conductor TEM mode transmission line with the cross-sectional configuration as shown in Fig. 10.1. It consists of a central strip with its width

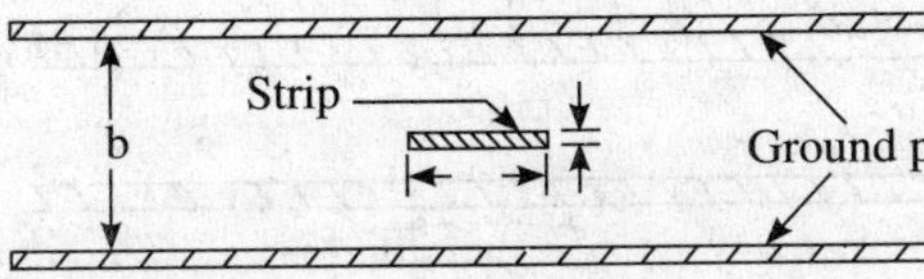

Fig. 10.1

much greater than its thickness ($W \gg t$) and two ground planes which extend considerably in transverse directions. The space between the ground planes is filled with a dielectric medium and the central strip is embedded in this dielectric. A practical embodiment of this configuration is shown in Fig. 10.2. This figure shows two laminates (*A* and *B*) used for obtaining the stripline geometry. The laminate *A* is completely metallized on the bottom side and has a strip of required width *W* on the other side. The laminate *B* has metallization on the top side and is clear of any metal on the bottom side. The two laminates are screwed (or clamped) together to yield the final configuration. As the thickness '*t*' is very small (1.4 mil to 2.8 mil) any air pocket existing after pressing two laminates together may be ignored.

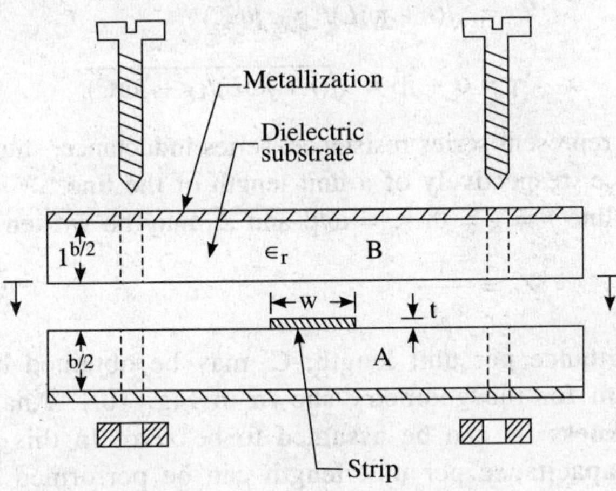

Fig. 10.2

The mode of propagation along a stripline is transverse electromagnetic (TEM). Field lines are concentrated near the central conductor. The ground planes are at zero potential. The electric and the magnetic field distributions are as sketched in Fig. 10.3 (a) and (b), respectively. As one moves in the mid-plane away from the strip, the field decays rapidly. This suggests that the geometry could be terminated in the transverse direction without affecting the transmission characteristics.

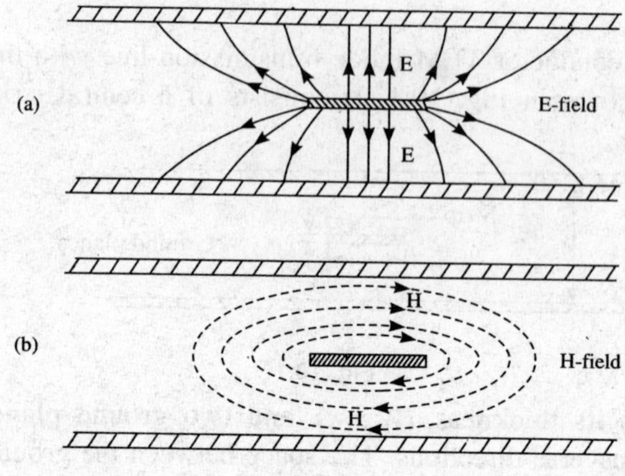

Fig. 10.3

As in the case of other TEM lines, the phase velocity v_p of the wave along the stripline is given by

$$v_p = c/\sqrt{\varepsilon_r} \quad ...(10.1)$$

where c is the velocity of electromagnetic waves in free space and ε_r is the relative permittivity of the dielectric medium. The transmission line parameters, characteristic impedance Z_0 and phase constant β, may be expressed in terms of line constants as follows:

$$Z_0 = \sqrt{(r + j\omega L)/(g + j\omega C)} \quad ...(10.2)$$

and

$$\gamma = \alpha + j\beta = \sqrt{(r + j\omega L)/(g + j\omega C)} \quad ...(10.3)$$

where r, L, g and C represent series resistance, series inductance, shunt conductance and shunt capacitance, respectively of a unit length of the line.

For a lossless line $r = g = 0$, $v_p = \omega/\beta$ and Z_0 may be written as

$$Z_0 = \frac{1}{v_p C} \quad ...(10.4)$$

The value of capacitance per unit length, C, may be obtained by solving the electrostatic problem for the geometry shown in Fig. 10.1. Analysis becomes simpler if the thickness 't' can be assumed to be zero. In this case an exact derivation of the capacitance per unit length can be performed by the use of conformal transformations. This formulation has been described by Cohn (1954) and the characteristic impedance is derived to be

$$Z_0\sqrt{\varepsilon_r} = 30\pi \frac{K(k)}{K'(k)} \text{ ohm} \quad ...(10.5)$$

where $k = \text{sech}\,(\pi W/2b)$ and K, K' represent elliptic integral and complementary elliptic integral, respectively and are defined as

$$K(k) = \int_0^{\pi/2} d\phi\sqrt{1 - k^2 \sin\phi} \quad ...(10.6)$$

$$K'(k) = K(k') \qquad \text{...(10.7)}$$

with $k' = (1 - k^2)^{1/2} = \tanh(\pi W/2b)$.

The ratio K/K' is approximated fairly accurately (3 to 8 parts per million) by

$$K/K' \simeq \frac{1}{\pi} \ln\left\{2\frac{1+\sqrt{k}}{1-\sqrt{k}}\right\} = F_1(k) \qquad \text{...(10.8)}$$

for $0.5 \le k^2 \le 1$; and for $0 \le k^2 \le 0.5$ we may write

$$K/K' \simeq 1/F_1(k') \qquad \text{...(10.9)}$$

Using these approximations, the equation (10.5) for characteristic impedance Z_0 may be written as

For $W/b \le 0.5$

$$Z_0\sqrt{\varepsilon_r} = 30 \ln\left\{2\left(\frac{1+\sqrt{k}}{1-\sqrt{k}}\right)\right\} \text{ ohms} \qquad \text{...(10.10)}$$

and for $W/b > 0.5$

$$Z_0\sqrt{\varepsilon_r} = 30\pi^2/\ln\{2[(1+\sqrt{k'})/(1-\sqrt{k'})]\} \text{ ohms} \qquad \text{...(10.11)}$$

These formulas yield results which are virtually exact.

When the strip thickness is not negligible, an exact analysis can still be carried out by using conformal transformation techniques, but with considerable difficulty and with the disadvantage that the resulting formulas are very complicated and yield the characteristic impedance implicitly rather than explicitly. There is extensive literature available on the attempts to derive accurate but simple

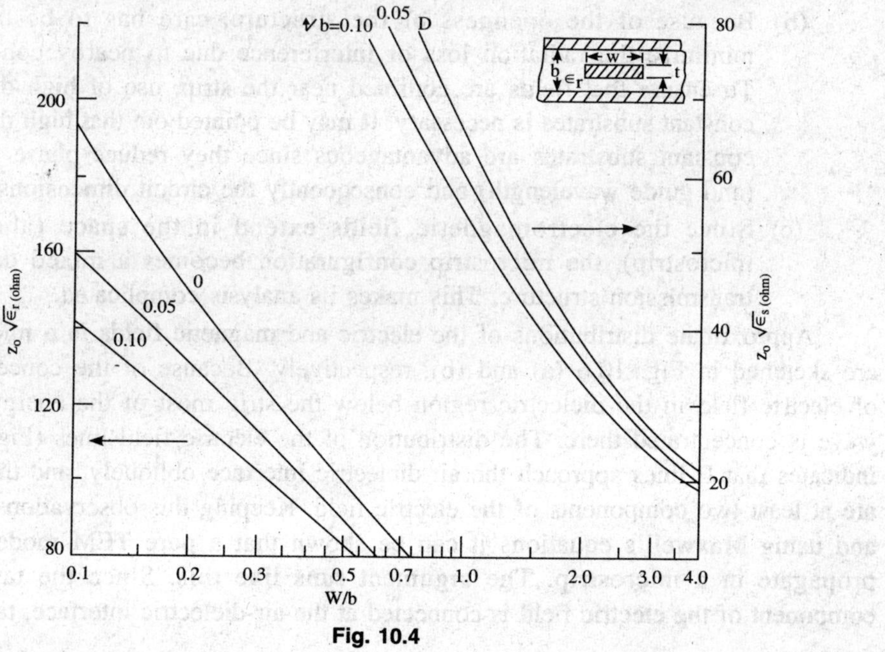

Fig. 10.4

approximate formulas. However, for microwave integrated circuits, fabricated by thin film technology, values of the thickness t used normally are about 1.4 or 2.8 mils, and errors introduced by assuming $t = 0$ are small. Figure 10.4 shows the variation of the characteristic impedance of striplines with geometry.

As in the case of coaxial line systems, it is usually necessary or desirable to ensure the non-propagation of higher order (non-TEM) modes. This can be achieved by ensuring that $b < \lambda_d/2$, where λ_d is the wavelength in the dielectric medium.

10.1.2 Microstrip

A cross-sectional view of a 'microstrip' is shown in Fig. 10.5. This line consists of a single dielectric substrate with ground plane on one side and a strip on the other face. The top ground plane of stripline configuration is not present in this case.

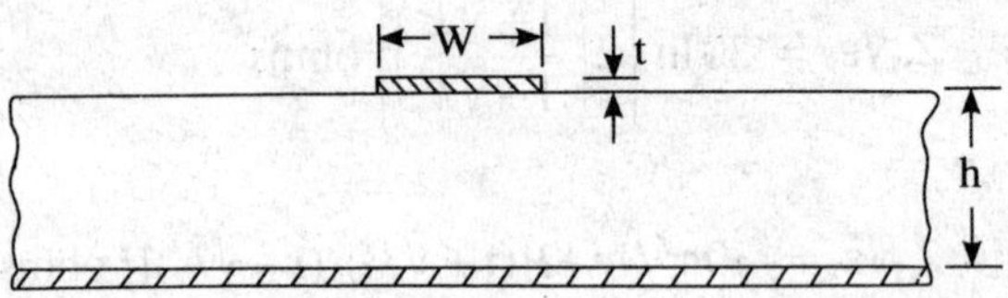

Fig. 10.5

This gives rise to the following feature of microstrip lines which are not available in striplines:

(a) There is an easy access to the top surface which makes it very convenient to mount discrete (active or passive) devices and to make minor adjustments after the circuit has been fabricated.

(b) Because of the openness of the structure, care has to be taken to minimize the radiation loss or interference due to nearby conductors. To ensure that fields are confined near the strip, use of high dielectric constant substrates is necessary. It may be pointed out that high dielectric constant substrates are advantageous since they reduce phase velocity (and guide wavelength) and consequently the circuit dimensions also.

(c) Since the electromagnetic fields extend in the space (above the microstrip), the microstrip configuration becomes a mixed dielectric transmission structure. This makes its analysis complicated.

Approximate distributions of the electric and magnetic fields in a micro-strip are sketched in Fig. 10.6 (a) and (b), respectively. Because of the concentration of electric field in the dielectric region below the strip most of the energy of the wave is concentrated there. The distribution of the electric field lines (Fig. 10.6a) indicates that E lines approach the air-dielectric interface obliquely, and thus there are at least two components of the electric field. Keeping this observation in mind and using Maxwell's equations it can be shown that a pure TEM mode cannot propagate in a microstrip. The argument runs like this. Since the tangential component of the electric field is connected at the air-dielectric interface, tangential

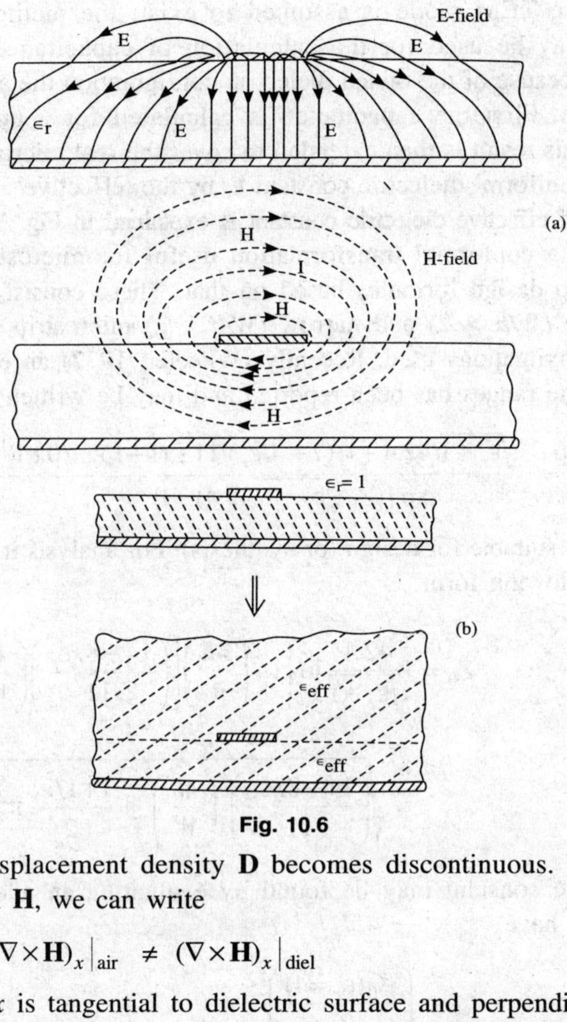

Fig. 10.6

component of displacement density **D** becomes discontinuous. Using Maxwell's equation for Curl **H**, we can write

$$(\nabla \times \mathbf{H})_x \big|_{\text{air}} \neq (\nabla \times \mathbf{H})_x \big|_{\text{diel}} \qquad \text{...(10.12)}$$

where direction x is tangential to dielectric surface and perpendicular to the strip conductor. If H_z is zero (i.e. if the mode is pure TEM), Eqn. (10.12) gives

$$\partial H_y \big|\partial z\big|_{\text{air}} \neq \partial H_y / \partial z\big|_{\text{diel}}$$

or

$$H_y \big|_{\text{air}} \neq H_y \big|_{\text{diel}} \qquad \text{...(10.13)}$$

Inequality in (10.13) violates the field matching conditions for the normal* **component of magnetic field. This implies that H_z should be a non-zero quantity for (10.12) to be satisfied. This leads to the conclusion that a pure TEM mode cannot be supported by a microstrip line. However, since the major portion of electric field lines is concentrated below the strip, the electric flux crossing the air dielectric boundary is small. Therefore, the deviation from the TEM mode is small and may be ignored for most of the circuit design applications.**

---

* y-direction is normal to the strip and the substrate and the propagation is along z.

When a pure TEM mode is assumed to exist, the method of conformal transformation may be used for the calculation of capacitance in the case of microstrip also. Because of the mixed dielectric configuration the analysis is carried out in two stages. First the capacitance is calculated for a uniform dielectric microstrip line. This result is then extended to cover the real microstrip case simply by replacing the 'uniform' dielectric constant ε_r by the 'effective' dielectric constant ε_{eff}. The concept of effective dielectric constant is explained in Fig. 10.6 (b). Wheeler (1965) has given a conformal transformation useful for microstrip analysis and derived micro-strip design formulas based on that. These consist of two different relations for wide *(W/h* > 2) and narrow *(W/h* < 2) microstrip lines because of the different approximations used. Recently (Wheeler, 1977) an empirical relation that covers both the ranges has been reported and may be written as

$$\frac{W}{h} = 8\frac{\{[\exp Z_0\sqrt{\varepsilon_r+1}/42.4-1](7+4/\varepsilon_r)/11+(1+1/\varepsilon_r)/0.81\}^{1/2}}{\exp\{(Z_0\sqrt{\varepsilon_r+1}/42.4)-1\}} \qquad ...(10.14)$$

Relation (10.14) is suitable for design (or synthesis). For analysis it may be reversed and put in the following form

$$Z_0 = \frac{42.4}{\sqrt{\varepsilon_r+1}}\ln\left\{1+\left(\frac{4h}{W}\right)\left[\left(\frac{14+8/\varepsilon_r}{11}\right)\left(\frac{4h}{W}\right) + \sqrt{\left(\frac{14+8/\varepsilon_r}{11}\right)^2\left(\frac{4h}{W}\right)^2+\frac{1+1/\varepsilon_r}{2}\pi^2}\right]\right\} \qquad ...(10.15)$$

Effective dielectric constant may be found by evaluating an additional value of Z_0 for $\varepsilon_r = 1$. We have

$$\varepsilon_{eff} = \left(\frac{Z_0(\varepsilon_r=1)}{Z_0}\right)^2 \qquad ...(10.16a)$$

Alternatively, an empirical expression (Schneider, 1969) gives ε_{eff} directly in terms of ε_r, *W* and *h*. This may be written as

$$\varepsilon_{eff} = \frac{\varepsilon_r+1}{2}+\frac{\varepsilon_r-1}{2}\left(1+\frac{10h}{W}\right)^{-1/2} \qquad ...(10.16b)$$

Variation of microstrip characteristic impedance with *W/h* is shown in Fig. 10.7. It may be noted that for microstrip lines, the phase velocity v_p is given by

$$v_p = c/\sqrt{\varepsilon_{eff}} \qquad ...(10.17)$$

Since ε_{eff} is a function of *W* and *h*, the phase velocity and hence the guide wavelength depends upon the line impedance also. This factor has to be taken into account while designing microstrip circuits.

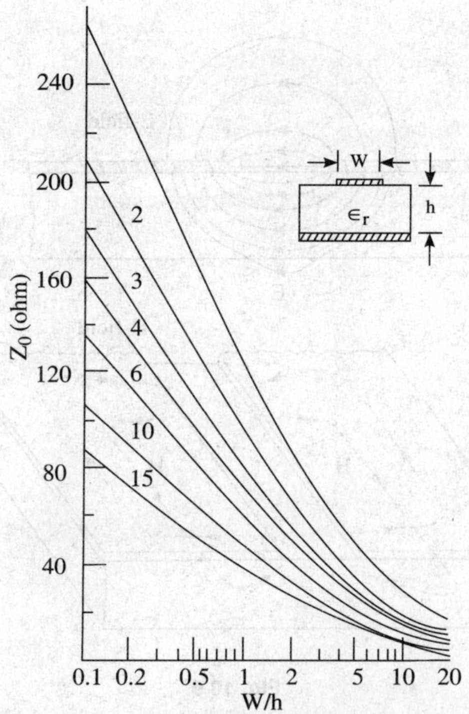

Fig. 10.7

10.1.3 Slot line

Although microstrip is the most commonly used transmission structure for microwave integrated circuits, a slot line (shown in Fig. 10.8) is an alternative transmission structure. Approximate electric and magnetic field distributions in this structure are shown in Figs. 10.9 (a) and (b), respectively. It may be observed that the magnetic field has a component in the direction of propagation also. Thus the mode of propagation is a transverse electric mode and not a TEM mode.

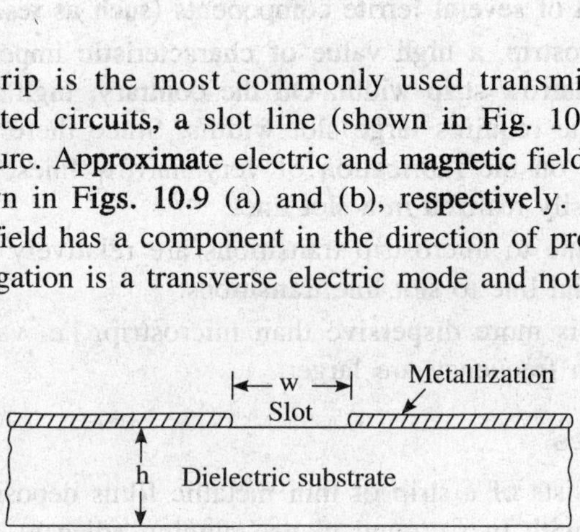

Fig. 10.8

Main features of a slot line vis-a-vis a microstrip are as follows :

(i) In a slot line, both the conductors are in one plane and therefore shunt mounting of components (active or passive) across the line is very convenient.

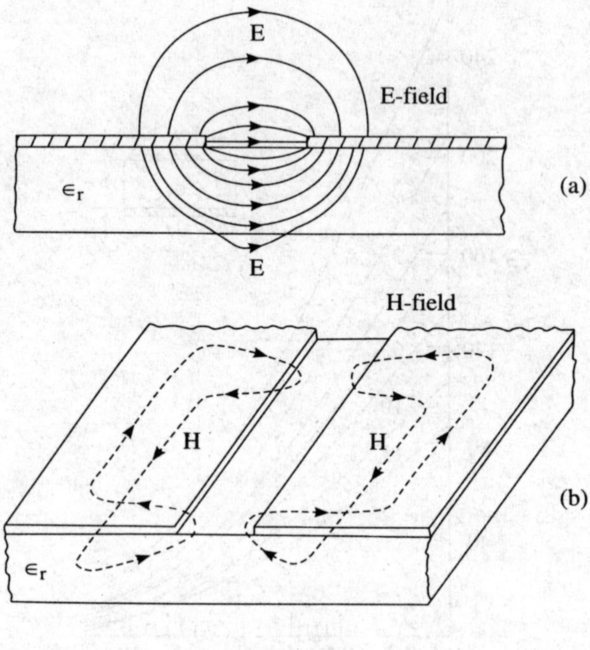

Fig. 10.9

(ii) We can find a location (in the cross-section of the slot line) where the magnetic field is circularly polarized. This characteristic is useful for the design of several ferrite components (such as resonance isolators).

(iii) In a microstrip, a high value of characteristic impedance is produced by very narrow strip width. On the contrary, high impedance in case of slot line requires large slot widths. Since there is a technological limitation on the fabrication of very narrow lines, higher impedance can be easily realized in a slot line.

(iv) Coaxial line to microstrip transitions are relatively easier to fabricate than coaxial line to slot line transitions.

(v) Slot line is more dispersive than microstrip, i.e. variations of Z_0 and λ_0/λ_g with frequency are larger.

10.1.4 Coplanar lines

A coplanar line consists of a strip of thin metallic films deposited on the surface of a dielectric slab with two ground planes running adjacent and parallel to the strip on the same side. This configuration is shown in Fig. 10.10. This geometry combines some of the advantages of microstrips and slot lines in a single structure. Some interesting features of coplanar lines are listed below.

(i) The microstrip geometry is convenient for series mounting of components—across a gap in the strip conductor. On the other hand, slot lines are well suited for shunt mounting of components—across the slot. In a coplanar line both series and shunt mounted components can be easily incorporated.

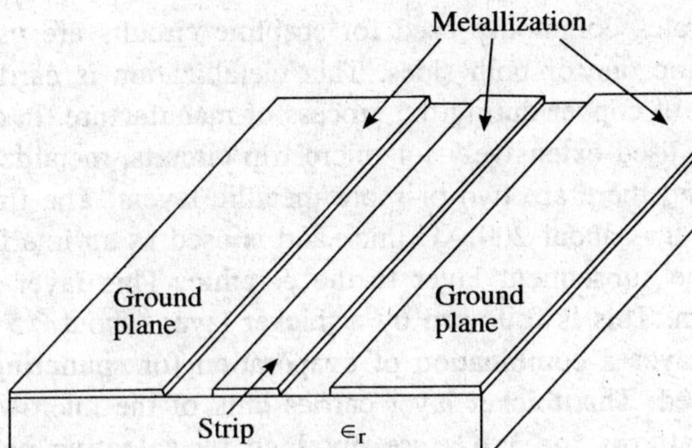

Fig. 10.10

(ii) Coaxial-to-coplanar line and microstrip-to-coplanar line transitions can be achieved easily.

(iii) Dimensions of a coplanar line can be tapered, for a uniform impedance, to connect a wide dimensioned component (for example an SMA connector) to a minute device chip or lumped element.

(iv) Both balanced and unbalanced modes are possible in a coplanar line. Thus this transmission line finds extensive application in balanced mixers, balanced modulators, etc; where both the balanced and unbalanced signals are present.

(v) Advantage of elliptically polarized magnetic field region obtainable for slot lines is present in the case of coplanar line also.

10.2 TECHNOLOGY OF HYBRID MICs

Preparation of an MIC starts with the selection of dielectric substrate. Choice of the substrate depends, to a certain extent, on the type of the transmission line (microstrip, stripline, slot line or coplanar line) to be used. For passive circuits which do not contain any discrete components (active or passive) one can use stripline configuration. On the other hand, if the circuit includes any active or discrete passive component, microstrip (or slot line, etc.) is more convenient. In order to avoid possibilities of radiation loss, microstrips (and other open lines) require substrates with higher dielectric constant than those used for stripline circuits. Commonly used dielectric substrates for microstrip circuits are alumina ($\varepsilon_r \simeq$ 8 to 10), sapphire ($\varepsilon_r \simeq$ 9.3 to 11.7), fused quartz ($\varepsilon_r \simeq$ 3.8) and beryllia ($\varepsilon_r \simeq$ 6.6). For stripline work, the commonly used substrates are polystyrene ($\varepsilon_r \simeq$

2.54), teflon ($\varepsilon_r \simeq$ 2.1) polyplefin ($\varepsilon_r \simeq$ 2.32) and many derivatives of these plastic materials.

Plastic substrates commonly used for stripline circuits are usually available with metallization on one or both sides. This metallization is carried out by hot-pressing a thin foil of copper during the process of manufacture. In case of ceramic or glass substrates, used extensively for microstrip circuits, metallization is carried out in two steps and there are two or more metallic layers. The first thin layer is usually chromium. It is about 200 AU thick and is used as an interface to improve the adhesion of the subsequent layer to the ceramic. This layer is obtained by vacuum evaporation. This is followed by a thicker layer (about 0.5 mil) of copper or gold. For this layer a combination of evaporation (or sputtering) followed by electroplating is used. This thicker layer carries bulk of the microwave current.

Conductor patterns for MICs are obtained by selective photo-etching of metallized substrate. First of all, a layout of the circuit is drawn and a photomask is prepared. The size of the mask is equal to the actual size of the circuit. This mask is a negative transparency, i.e. transparent at the portions where metallization is required. The substrate is cleaned and a thin layer of photoresist emulsion spread evenly over it. This layer is exposed to the ultraviolet light through the circuit mask. Exposed portions, corresponding to circuit pattern, harden up. Unexposed photoresist is dissolved away and etching of the metal (copper or gold and chromium) is carried out. Required metallization pattern is protected by hardened photoresist and thus left unetched on the substrate. This photoresist layer on the final circuit pattern is removed chemically.

This procedure is similar to that used in integrated circuit technology and is suitable for large-scale batch production. In this method of circuit fabrication, all the dimensions in MICs are defined by photo-etching and not by machining (by turning, milling, etc.) as is the case with conventional microwave circuits.

Next step after photo-etching of metallized substrate to obtain the required circuit pattern is to mount the discrete (active or passive) devices. The mounting of series-connected devices may be easily achieved by bonding the device across a gap in the microstrip. Wire-bonded device chips, leadless inverted devices (LIDs), ceramic carriers or beam-lead devices are well suited for MICs. Shunt mounting on the microstrip circuits is carried out by drilling a hole through the substrate (usually by an ultrasonic grinder). The device is mounted between the ground plane and a strap placed over the microstrip to bridge the hole. Devices in 'pill' or 'micropill' packages are well suited for this application. Shunt mounting of devices in a slot line circuit is fairly easy. The device is mounted across the slot. Device packages that are suitable for series mounting in a microstrip are also equally suited for shunt mounting in slot line circuits. Bonding techniques for connecting device leads to microstrips, etc, depend on the circuit metallization and type of the device package used. Generally soldering is used with the copper clad laminates (used mainly for stripline circuits) and thermo-compression bonding is used for gold metallization (used for microstrip and slot line circuits). Low-temperature indium solders and conductive epoxies may also be used. Increasing use of chips and beam lead devices has led to the general acceptance of gold metallization.

Fabrication of lumped elements for MIC is discussed later in Chapter 11.

10.3 ADVANTAGES OF MICs

MICs are preferred over the conventional microwave circuits for the following reasons: (i) small size and weight. (ii) reproducibility and reliability, (iii) improved performance in some cases, and (iv) eventual cost reduction when producd in large quantities.

10.3.1 Small size and weight

Miniaturization is a key feature of MICs. Following factors contribute to make MICs smaller and lighter.

a. *Reduced cross-section for transmission lines:* An *X*-band waveguide is 1 inch by 0.5 inch size whereas cross-section area required for a 50 ohm microstrip on alumina substrate is of the order of 0.25 inch by 0.25 inch. Microstrip dimensions are smaller than even the standard coaxial lines.

b. *Reduced guide wavelength:* Use of high dielectric constant substrates reduces the guide wavelength by a factor equal to the square-root of effective dielectric constant. For alumina circuits a reduction of 2.5 to 3.0 is obtained. This reduces the size of circuits using distributed elements in the same ratio.

c. *Use of lumped elements:* A further reduction in the size is achieved by the use of lumped elements in place of distributed circuits. Design and usage of lumped elements is discussed in details in Chapter 11. Dimensions of lumped elements are necessarily less than about one-tenth of the wavelength. Thus the lumped element circuits are about an order magnitude smaller than distributed circuits.

d. *Elimination of interconnections and flexibility in layout :* Since complicated circuit functions cannot be built in single unit when waveguides or coaxial lines are used, a conventional subsystem consists of number of separate circuits connected together by coaxial connectors, waveguide flanges or adapters. In MICs even a large complicated circuit function can be produced by photo-etching of a single metallization and a considerable volume occupied by connectors, etc. is saved. Also this fabrication process allows the designer to locate the input-output ports at the desired location. Thus the amount of hardware, which conventional microwave circuits use for bends etc., is saved.

e. *Use of device chips :* In MICs it is possible to use uncapsulated device chips. These device chips occupy much smaller space than packaged devices. Also, the mounting structure for packaged devices may require additional space.

10.3.2 Reproducibility and reliability

Techniques of MIC fabrication, i.e. thin film deposition and photo-etching, lead to a very uniform product. Variations are caused by tolerances in semiconductor device characteristics and substrate material. Methods to reduce or compensate for these variables are being developed as MIC technology matures. Increased reliablity of MICs stems from the following factors.

(i) Microwave semiconductor devices used in MICs have longer life as compared to that of tubes.

(ii) A fewer number of connectors are required for achieving a given subsystem function.

(iii) More durable connection between device and circuit is obtained because of improved bonding techniques. Possibility of relative motion among various parts of the circuit is little. Therefore, the circuit construction is more rigid.

10.3.3 Improvement in performance

In several situations, the circuit performance may be improved by adopting MIC technique. This contention is supported by the following observations.

a. Improvement in device-circuit interface: In conventional microwave circuits, the transverse dimensions of the circuit are much larger than those for the semiconductor devices used. This necessitates the use of mounting structures, like posts, etc. which introduce additional reactances and limit the circuit performance. In MICs, transverse dimensions of the microwave structures are compatible with the size of device chips. The package and the mounting reactances are eliminated. Elimination of these reactances increases the circuit bandwidth and also the device impedance level. Ultrabroadband PIN diode switches have been obtained in stripline and microstrip configurations only.

b. Use of lumped elements : Lumped element resonators (as discussed later in Sec. 11.4) have a wider bandwidth than the corresponding distributed element circuits. This is caused by the fact that the equivalent L and C of the distributed circuit are not constant with frequency. Thus, the use of lumped inductors allows one to have, for example, a wider bandwidth in tunnel diode amplifiers and an increased tuning range in varactor tuned circuits.

c. Use of special circuits combining microstrip and other lines : Combination of microstrip and slot line in the same circuit introduces another degree of flexibility in circuit design. An increase in bandwidth of couplers (as large as ten octaves!) has been reported by using such a combination. Similarly, slot line and coplanar lines have been combined usefully to design novel circuit configurations for balanced mixers.

10.3.4 Reduction in cost

For production in large quantities, MIC technology promises a reduction in the circuit cost by almost an order of magnitude. Various factors that lead to the reduced cost may be listed as follows :

(i) Comparing MICs with the conventional microwave circuits, one notes that time-consuming and expensive operation of precision machining has been replaced by photoetching which is a much quicker and cheaper process. As an example, let us consider a reported comparison of the two techniques at a research laboratory stage. A conventional tunnel diode amplifier is built in six man-weeks. Corresponding lumped element MICs have been built at the rate of six per day.

(ii) The cost of fabrication of MICs is, to a considerable extent, independent of the circuit complexity. For example, etching out a microstrip section or filter circuit requires about the same effort.

(iii) Once the circuit has been perfected and mask prepared, large-scale production does not require any additional tooling, etc.

(iv) MICs can use semiconductor devices in chip form. Chips are much cheaper than the packaged devices.

10.4 DIFFICULTIES WITH MICs

There are some difficulties associated with the design and usage of MICs. Most of these are because of the fact that once an MIC has been fabricated, there is hardly any provision for adjustment of circuit parameters. It is very difficult to include arrangements like tuning screws, variable shorts, etc. Nor are these arrangements desirable from the point of view of reliability and reproducibility. Consequently, circuit design philosophy has to be different from that used for conventional microwave circuits. Since screws and plungers cannot be incorporated to improve the circuit performance, the circuits have to be designed fairly accurately taking all the parameters into account so that the adjustments required after fabrication are reduced to minimum.

In order to design the circuits accurately, precise characterization of semiconductor devices is needed. This poses considerable problems, specially when the devices used are in chip form or miniature packages.

Consequent to the considerations mentioned above, computer-aided-design and computer-aided-measurement techniques are well suited for the design of MICs. These allow all the tolerances, parameters and their variations to be taken into account and to analyse their effect on the performance even before the circuit is fabricated.

One disadvantage of MICs is the low values of the Q factor associated with microstrip resonators (of the order of 100) as compared with waveguide resonators (of the order of 1000). Solid state oscillators with high frequency stability are difficult to obtain in microstrip configuration. However, dielectric disk resonators provide high Q values and may be used in MICs.

Power handling capacity of MICs is of course much less than that of waveguide circuits. However, 10W power amplifiers at 1.5 GHz have been developed commercially.

SELECT BIBLIOGRAPHY

K.C. Gupta and A. Singh, *Microwave Integrated Circuits,* Wiley Eastern, New Delhi (1974). Apart from discussions on various aspects of MICs this book contains an extensive bibliography of papers on MICs published during 1967 to 1973.

M.A.R. Gunston, *Microwave Transmission Line Impedance Data,* Van Nostrand Reinhold Company, London (1972).

H. Howe, *Stripline Circuit Design,* Artech House, 1974.

M.V. Schneider, 'Microstrip lines for microwave integrated circuits', *B.S.T. Journal,* Vol. 48, (No. 5), pp. 1421-44 (1969).

H. Sobol, 'Applications of integrated circuit technology to microwave frequencies', Proc. IEEE, 59, p. 1200(1971).

H.A. Wheeler, 'Transmission-line properties of a strip on a dielectric sheet on a plane' *IEEE Trans.,* MTT-25, p. 631-647 (1977).

------- 'Transmission-line properties of parallel strips separated by a dielectric sheet', *IEEE Trans.,* MTT-13, pp. 172-185 (1965).

K.C. Gupta, Ramesh Garg and I.J. Bahl, *'Microstrip Lines and Slotlines'* Artech House (Mass., U.S.A.) 1979.

PROBLEMS

10.1 Equations (10.10) and (10.11) give values of characteristic impedance for striplines in terms of various dimensions. Often in design problems it becomes necessary to calculate strip width to substrate height ratio for a given impedance. Rearrange eqns. (10.10) and (10.11) to give W/b in terms of Z_0.

10.2 A 10-ohm resistive load is to be matched to a 100-ohm stripline. Design a 2-section maximally flat quarter-wave transformer. Specify the final circuit layout when a 1/16 inch thick polystyrene substrate ($\varepsilon_r = 2.55$) is used for stripline circuit. Operating frequency is 3.0 GHz. Expressions for W/b derived in Prob. 10.1 may be used.

10.3 Derive the relation for microstrip analysis (eqn. 10.15) from that for micro-strip design (eqn. 10.14).

10.4 A 50-ohm microstrip line is to be matched to a 100-ohm resistive load. Find the width and length of a single section quarter-wave transformer. The design employs a 0.025 inch thick alumina ($\varepsilon_r = 9.7$) substrate and the operating frequency is 10 GHz.

10.5 Design a half-wave resonator at 3 GHz using a 50-ohm microstrip on 0.025 inch alumina (ε_r = 9.7) substrate. Fringing capacitance at each open end of the line is 0.1 pF. Calculate the length of microstrip line required for resonator (Smith chart may be used).

10.6 Microstrip ring resonators are frequently used for measurement of microstrip phase velocity (and hence guide wavelength). A ring resonator with 1 cm mean diameter is fabricated on alumina substrate (ε_r = 10.0, thickness = 0.025 inch). If the width of the microstrip is also 0.025 inch, calculate first three lowest resonance frequencies for the ring. (Modification in phase velocity because of the curvature of the microstrip may be ignored.)

10.7 A 30 pF lumped thin film capacitor is used as a DC block in a microstrip circuit on alumina substrate (ε_r = 10.0, thickness = 0.025 inch). If the microstrip impedance is 50 ohms, calculate the insertion loss that will be introduced at 3.0 GHz because of such a DC block.

10.8 A 50-ohm microstripline is fabricated on a 0.025 inch thick alumina (ε_r = 10.0) substrate. Value of ε_r can vary from substrate to substrate by 2 per cent. Calculate the corresponding fluctuations in impedance and guide wavelength at 3 GHz.

10.9 A 50-ohm stripline is fabricated on polystyrene substrate (ε_r = 2.54) and b = 0.125 inch (*see* Fig. 10.1). If the accuracy that can be achieved in photoetching process is limited to ± 0.001 inch, calculate the corresponding fluctuation in the line impedance.

10.10 Sketch roughly the electric and magnetic field distribution in the cross-section of a coplanar line. (Assume central strip at a high potential and outer conductors at ground potential for plotting electric field.)

CHAPTER 11

Lumped Elements at Microwave Frequencies

In conventional microwave circuits, distributed elements employing sections of transmission lines or waveguides are used. This is so because the size of normally used lumped elements (resistors, inductors and capacitors) becomes comparable to the wavelength at microwave frequencies. However, if the size of lumped elements can be reduced, they can be used at these frequencies also. Lumped elements for use at microwave frequencies are designed on the basis of this consideration. With the advent of photolithography and thin film techniques, the size of the elements can be reduced such that these elements may be operated up to *X*-band frequencies. In this chapter some features of design and applications of these lumped elements are discussed.

11.1 DESIGN OF LUMPED ELEMENTS

Design of lumped elements at microwave frequencies may be arrived at by considering them as very short sections (much smaller than wavelength) of TEM transmission lines. Consider a transmission line with series inductance L per unit length, series resistance r per unit length, shunt capacitance C per unit length and shunt conductance g per unit length. By using transmission line formula discussed in Chapter 2 (eqn. 2.31), the input impedance of a length l of a transmission line terminated in an impedance Z_L is written as

$$Z_{\text{in}} = Z_0 \left\{ \frac{Z_L \cosh \gamma l + Z_0 \sinh \gamma l}{Z_0 \cosh \gamma l + Z_L \sinh \gamma l} \right\} \quad ...(11.1)$$

where Z_0 is the characteristic impedance of the transmission line and is given by

$$Z_0 = \sqrt{(r + j\omega L)/(g + j\omega C)} \quad ...(11.2)$$

and γ is the propagation constant given by

$$\gamma = \sqrt{(r + j\omega L)(g + j\omega C)} \quad ...(11.3)$$

when $\gamma l << 1$, Z_{in} may be written as

$$Z_{\text{in}} = Z_0 \frac{Z_L + Z_0 \gamma l}{Z_0 + Z_L \gamma l} \quad ...(11.4)$$

11.1.1 Resistors and inductors

When a small section ($\gamma l << 1$) of a transmission line is terminated in a short circuit, the input impedance may be written as

$$Z_{in} = Z_0 \gamma l = (r + j\omega L)\, l \qquad ...(11.5)$$

The two-terminal circuit element thus obtained behaves like a resistor or an inductor depending on the value of r as compared to that of ωL. However, it is not necessary to use both the conductors of a transmission line to obtain a resistor or an inductor. A single metallic strip on a dielectric substrate is sufficient for this purpose. In this case, values of r and L pertain to this metallic strip. As before, r is the resistance (at microwave frequencies) per unit length and L is the self-inductance per unit length. Terman (1943) has given the formulas for r and L for a ribbon of rectangular cross-section. The value of r may be calculated from the equation

$$R_{ac} = rl = \frac{K R_s}{2(W+t)} l \qquad ...(11.6)$$

where R_s is the sheet resistivity* in ohms per square, W the width and t is the thickness of metallic ribbon. K is a correction factor that takes into account the crowding of current at the corners of the ribbon. Value of K is a function of W/t and varies from about 1.3 to 2 as W/t varies from 1 to 100. The value of K is about 1.5 for a typical W/t = 10.

Inductance of a rectangular ribbon is given by

$$Ll \simeq 5.08\times10^{-3} l \left\{ \ln\frac{l}{W+t} + 1.193 + 0.2235 \frac{W+t}{l} \right\} \qquad ...(11.7)$$

where Ll is in nanohenries and all dimensions are in mils (10^{-3} inch).

The unloaded Q of the ribbon inductor may be written as

$$Q = \omega L / R_{ac} \qquad ...(11.8)$$

For a copper ribbon ($R_s = 2.61 \times 10^{-7} f^{1/2}$) and for W/t = 10, one obtains

$$Q = 5.3 \times 10^7 f^{1/2}\, WL/K \qquad ...(11.9)$$

For these equations to hold good, thickness t must be appreciably greater than twice the skin depth. It is found that values of Q greater than 100 can be obtained for wide ribbons (l/W < 15) with an inductance of 1 nanohenry or greater at about 1 GHz. The ribbon will be practicable for inductances of 2 nanohenries or less. It would have to be only 110 mil long (with l/W = 10) in order to have a good Q. This length is less than one turn of 40 mil diameter spiral coil. For larger inductances, many spiral turns are necessary and the above expressions are not valid.

An approximate expression for the inductance of a spiral inductor (Fig. 11.1) with n turns may be written as

$$L_s = \frac{a^2 n^2}{8a + 11c} (nH) \qquad ...(11.10)$$

* Sheet resistivity (or surface resistivity) is expressed in ohms per square, since it is the resistance of a surface with equal length and width and is independent of the actual size.

where all the dimensions are in mils, $a = (d_0 + d_i)/4$ (i.e. the average radius), $c = (d_0 - d_i)/2$, and d_i and d_0 are minimum and maximum diameters, respectively. The resistance of the spiral inductor may be written as

$$R = K' n\pi a\, R_s/W \qquad ...(11.11)$$

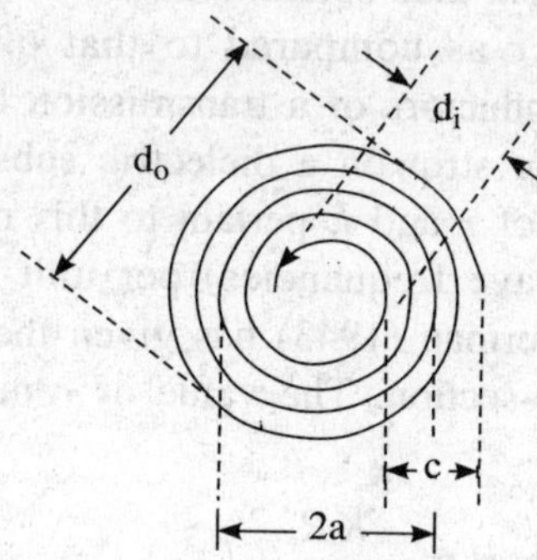

Fig. 11.1

where K' is a correction factor, similar to K for straight ribbon, and accounts for crowding of the current. From the above two relations, Q for the spiral coil may be written as

$$Q = \frac{2\times 10^{-9} fW na}{K' R_s (8a + 11c)} \qquad ...(11.12)$$

An assigned value of $K' = 1.5$ gives resonable agreement between designed and measured value of Q. For these spiral inductors Q's of about 100 have been achieved. Some salient considerations useful in the design of these inductors may be summarized as follows:

1. The spiral should have the widest ribbon possible (W large) while keeping the overall diameter (d_0) small. This implies that separation S between the turns should be as small as possible.
2. There should be some space at the centre of spiral to allow the flux lines to pass through (increasing the stored energy per unit length). It has been reported that $d_i/d_0 = 1/5$ yields optimum values of Q.
3. The Q of the inductors (as given by 11.12) also increases with the square root of frequency. This can be seen from equation (11.12) keeping in mind that R_s increases as square-root of frequency.
4. It is observed that for the same value of d_0, the Q of a circular spiral is higher by about 10 per cent than that of a square spiral.
5. Multi-turn coils have a higher Q due to higher inductance per unit area, but because of the presence of inter-turn capacitance they have lower self-resonance frequencies.

11.1.2 Capacitors

A lumped capacitor may be visualized as a small length of an open-circuited line for which input impedance may be written as

$$Z_{in} = Z_0/\tanh(\gamma l) \quad \text{...(11.13)}$$

When $\gamma l \ll 1$, and $g \ll \omega C$ equation (11.13) may be simplified as

$$Z_{in} \simeq \frac{Z_0}{\gamma l} = \frac{1}{gl + j\omega Cl} = \frac{g}{\omega^2 C^2 l} + \frac{1}{j\omega Cl} \quad \text{...(11.14)}$$

Its equivalent circuit will thus be a resistance $g/(\omega^2 C^2 l)$ in series with the capacitor Cl. A second order approximation yields (ignoring L):

$$Z_{in} = \frac{Z_0}{\gamma l} + \frac{Z_0 \gamma l}{3} = \frac{g}{\omega^2 C^2 l} + \frac{1}{j\omega Cl} + \frac{rl}{3} \quad \text{...(11.15)}$$

This implies an equivalent circuit of two resistors and a capacitor in series as shown in Fig. 11.2. When rl is ignored, this reduces to the equivalent circuit given by the equation (11.14). The term $(rl/3)$ represents the conductor loss and may be written as

$$\frac{rl}{3} = \frac{2R_s}{W}\frac{l}{3} \quad \text{...(11.16)}$$

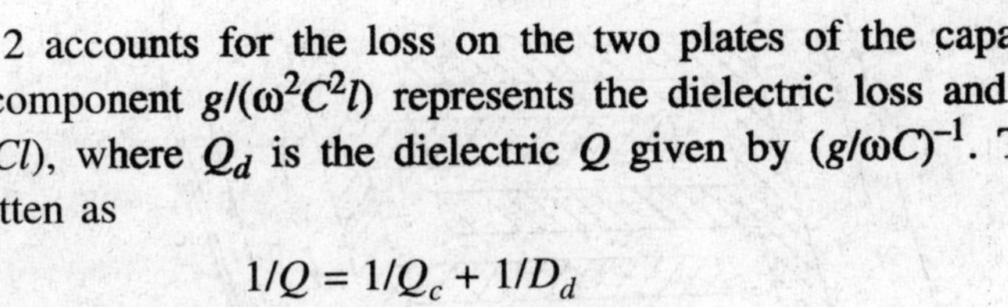

Fig. 11.2

The factor 2 accounts for the loss on the two plates of the capacitor. The other resistance component $g/(\omega^2 C^2 l)$ represents the dielectric loss and may be written as $1/(Q_d\, \omega Cl)$, where Q_d is the dielectric Q given by $(g/\omega C)^{-1}$. The total Q may thus be written as

$$1/Q = 1/Q_c + 1/D_d \quad \text{...(11.17)}$$

where the conductor Q_c is

$$Q_c = \frac{1}{\omega Cl}\frac{3W}{2R_s l} \quad \text{...(11.18)}$$

For a square shaped capacitor (i.e. $W = l$) with copper plates, and of total capacitance C_T (in pF) equation (11.18) simplifies to

$$Q_c = \frac{2.9 \times 10^4}{f^{3/2}(\text{GHz})\, C_T(\text{pF})} \quad \text{...(11.19)}$$

SANDWICH TYPE CAPACITORS

In order to have a large value of capacitance per unit area, it is desirable to have a small spacing between the two conductors (top and bottom) of the capacitor. This is achieved by fabricating a metal-dielectric-metal structure on the top surface of the substrate. The basic geometry of the sandwich capacitor is shown in Fig. 11.3. The dielectric used for this sandwich (normally silicon dioxide or silicon monoxide) is typically 0.02 to 0.04 mil (0.5 to 1.0 μm) thick. The technique of fabrication is discussed in the next section. Capacitance values in tens of picofarads are possible. This type of capacitors have been fabricated with dielectric Q of 600 at 1 GHz. Q at higher frequencies is lower in accordance with equation (11.19).

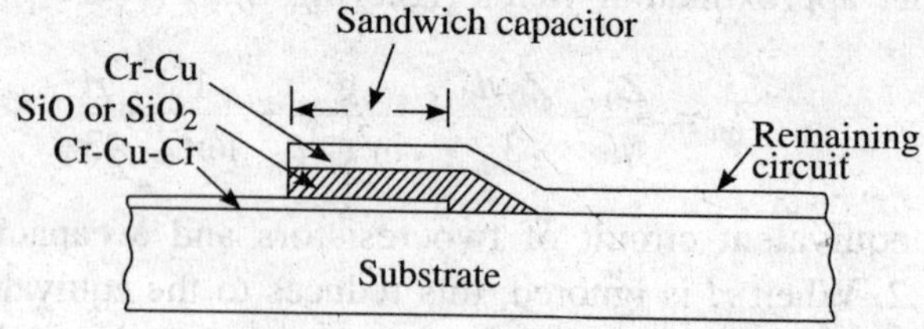

Fig. 11.3

INTERDIGITAL CAPACITORS

An alternative to this sandwich geometry is the interdigital capacitor shown in Fig. 11.4. In this geometry, both the conductors of the capacitor are in the same plane which is the top surface of the dielectric substrate used. This type of capacitors can be fabricated by the technique used in the fabrication of conventional microstrip circuits and does not require any additional processing steps when fabricated with microstrip circuits.

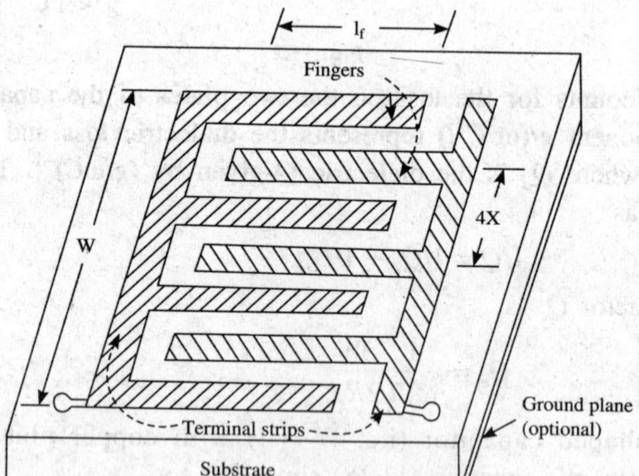

Fig. 11.4

An analysis for interdigital capacitor has been reported by Alley (1970). The capacitance between the two sets of digits in interdigital structure is found by using the capacitance formula for coupled microstrip lines. As in the case of sandwich geometry, both the dielectric loss and the conductor loss determine Q. Order of Q obtainable is indicated by the following reported results. The Q of a 2.83 pF capacitor ($N = 40$, $l_f = 44.8$ mil, $X = 2$ mil) fabricated on a 25 mil thick alumina substrate (99.5 per cent purity) has been measured to be 677 at 1.92 GHz.

Because of the inductance contributed by the equivalent transmission line, the interdigital capacitors show multiple resonances. Change in the capacitance value with frequency depends on the ratio W/l_f (*see* Fig. 11.4). The optimum value of W/l_f may be found numerically such that the capacitance variation with frequency is minimum. As an example, $W/l_f = 1.66$ and $N = 24$ makes the capacitance value constant up to a frequency of 3 GHz.

Interdigital capacitors with Q's comparable with those for sandwich type have been fabricated directly on alumina and no additional dielectric layer is required. They behave as lumped components up to 3 GHz and values from 0.1 to 10 pF can be achieved. The interdigital approach for capacitors requires a relatively large area (500 by 70 mil for a 12 pF value as compared with 22 × 22 mil in case of 1 μm thick SiO_2 capacitor). However, the interdigital capacitors are preferred because only a single layer of metallization is required.

11.1.3 Resonators

Lumped resonators at microwave frequencies consist of a suitable combination of a lumped inductor and a lumped capacitor. When interdigital capacitors are used the resonator fabrication involves only a single step. Both series and parallel resonant circuits are possible and have been used in filter circuits. The two configurations are shown in Fig. 11.5. It has been demonstrated that lumped-element parallel resonant circuits with single-turn inductors can be fabricated for operation from 5 to 10 GHz with associated resistance values of about 1 ohm or less. This corresponds to Q values between 10 and 90.

In addition to the *L-C* resonators mentioned above the YIG resonators and dielectric resonators could also possibly be incorporated in lumped element MICs.

11.2 FABRICATION OF LUMPED ELEMENTS

Techniques for fabricating lumped elements are similar to those used for microstrip circuits. Fabrication of inductors and interdigital type capacitors does not require any additional fabrication step when compared to the distributed microstrip circuits. The mask used for photoetching the metal-clad substrates includes the layouts for these lumped elements also. However, the sandwich type capacitors require a few additional steps in fabrication which would be discussed now.

The schematic in Fig. 11.3 illustrates the metal-dielectric-metal structure used for the sandwich type capacitors on a dielectric substrate. An alumina or glass substrate is normally used. The first step consists in depositing layers of Cr-Cu-

Cr (or Cr-Au-Cr) in a single vacuum evaporation. Chromium layers are very thin and are used for the adhesion of metal to the dielectric. Alternative metal combinations such as Ti-Au-Ti may also be used for this step. Cr-Au can introduce problems because of the very high resistance that often develops due to the diffusion of the two metals at high temperatures. The next step is a photoresist etch for defining the bottom layer of sandwich type capacitors. Metal layer is etched clean over the remaining portion of the substrate.

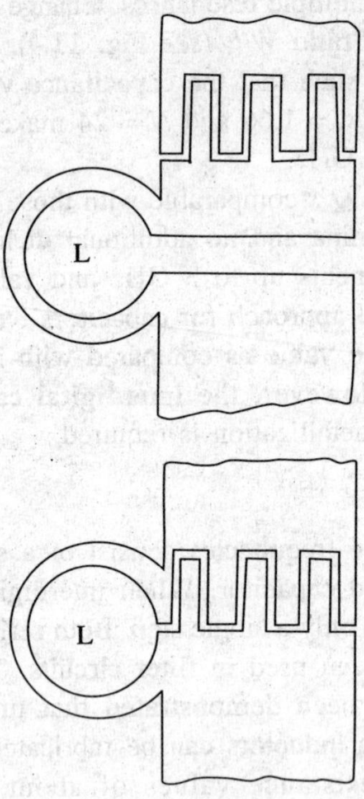

Fig. 11.5

The deposition of a silica (SiO_2) film is the next step. The silica film is grown by pyrolytic decomposition of silane. Silica layers so obtained have voltage breakdowns in excess of 10^7 V/cm, can be etched precisely and have low dielectric loss when densified. The thickness of SiO_2 film is of the order of 0.02 to 0.04 mil (05 to 1.0 μm). Silica from unwanted portions of the substrate is removed by selective photoetching. The oxide is then densified by heating to temperatures over 400°C. The effect of densification on Q is as follows : Q factor for deposited thin film SiO_2 is in the range of 20 to 80. It increases to about 600 when densified to 450°C. By densifying at 600°C, higher values of Q (greater than 5000) can be obtained. For comparison it may be noted that Q for fused quartz is about 10,000. Silicon monoxide (SiO) is also used as a dielectric material for sandwich capacitors.

A thin layer of top metal (typically 0.04 to 0.08 mil) is deposited on the already fabricated combination of metal and SiO_2 films. The remaining thickness of metal required for inductors and the rest of the circuit is selectively electroplated. The thin, unplated areas are etched clean. This technique is compatible with batch processing. It may be noted that the inductor has to be clear of the metal underneath. Resistive material can be incorporated as part of the adhesive layer.

11.3 MEASUREMENTS ON LUMPED ELEMENTS

Because of the small size of the lumped elements, special techniques are needed for measurements on these elements.

11.3.1 Slotted line measurements

Slotted line mesurements have been used for determining values of reactance. In this method, a 50-ohm microstrip line is terminated with the component under test. This fixture is connected through a sub-miniature adapter to a precision slotted line. The VSWR and the position of the minimum are recorded and subsequently repeated with the test component replaced by a known termination : a short or an open circuit. The reactance and the quality factor Q of the test component are determined by a comparison of the data for three cases. To obtain accurate results the circuit length between the slotted line and the component should meet the following requirements : (a) it should be well matched, (b) its loss should be small compared to the loss in test component, and (c) reactive end-effect at the terminal point on the transmission line where the test component is attached should be negligible.

Use of microstrip line for mounting the element is quite convenient. Connecting joints between the test piece and the line are accessible with the lumped element mounted at the end of the line. Mounting the test piece at the end of the 50-ohm line simplifies interpretation of the data. The 50-ohm microstrip line used has physical dimensions which are compatible with both the piece and sub-miniature connector. The distance between the strip conductor and the ground plane is sufficiently small (25 mil or less) to make the end effect negligible. The set-up has been used for measurements over the frequency range 0.5 to 2.5 GHz.

For evaluating inductors and capacitors, very high VSWRs are to be measured and the double minimum (or width of minimum) method is used. In this method

$$\text{VSWR} = \lambda/(\pi\Delta) \quad \text{...(11.20)}$$

where Δ is width of minimum at the points where the power is twice the minimum value.

11.3.2 Resonant line method

There is another method suggested for impedance measurement of microwave lumped elements that can be used up to 12 GHz. This technique is based on measurements of resonance frequency and Q of a resonant transmission line.

By connecting a lumped element to the line, the frequency of resonance and Q get perturbed, and from the perturbations, the impedance of the element can be found. Since the transmission line resonates at multiples of the fundamental frequency, the impedance of the lumped element can be found as a function of frequency.

Consider a transmission line open at one end and connected to a lumped element at the other end as shown in Fig. 11.6. This combination will resonate when

$$X = Z_0 \cot \theta \qquad \text{...(11.21)}$$

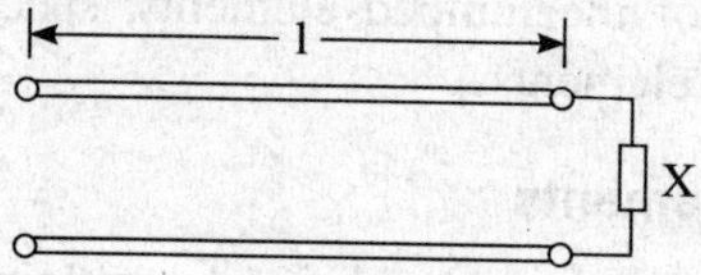

Fig. 11.6

where X is the reactive part of the impedance of the lumped element and θ is the electrical length of the transmission line given by

$$\theta = 2\pi f\, l/v \qquad \text{...(11.22)}$$

where f, l and v are the frequency of resonance, the length of the line and the velocity of wave propagation, respectively. Without the lumped element attached, this transmission line will resonate when the following condition is satisfied

$$2\pi f_n\, l/v = n\pi \qquad \text{...(11.23)}$$

when f_n is the nth order resonance frequency. From equations (11.22) and (11.23) we can find θ in terms of two resonance frequencies as

$$\theta = n\pi f / f_n \qquad \text{...(11.24)}$$

Combining this equation with (11.21) gives

$$X = Z_0 \cot (n\pi f / f_n) \qquad \text{...(11.25)}$$

where f is the resonance frequency for the circuit shown in Fig. 11.6 and f_n is the resonant frequency for the same line length with both ends open.

In a practical set-up, the value of f_n is measured first. Then the lumped element is attached and resonant frequency f measured. Alternatively, two identical lines can be fabricated on a single substrate and the lumped element attached to one of them. With this arrangement multiple resonance frequencies for the two lines can be measured and the value of reactance obtained over a range of discrete frequencies.

The Q factor of the lumped component can also be evaluated from the measurement of Q for unloaded line and that for the line loaded with the component. Since total power loss is equal to the sum of the power losses in the line and the element, we can write

$$\frac{U_{01}}{Q_{01}} = \frac{U_0}{Q_0} + \frac{U_1}{Q_1} \qquad \text{...(11.26)}$$

where U_{01}, U_0 and U_1 are the energies stored in the combined system, the unloaded line and the element, respectively, and Q_{01}, Q_0 and Q_1 are respective Q factors. Q_{01} is the sum of U_0 and U_1. Values of Q_{01} and Q_0 are measured experimentally. The energies U_1 and U_0 are calculated as follows. The energy stored in the transmission line, U_0, is given by

$$U_0 = \frac{1}{2}\frac{L_0}{\beta}\int_0^{\theta}[I_0 \sin(\beta x)]^2\, d(\beta x)$$

$$= \frac{1}{8}\frac{Z_0}{\omega} I_0^2 (2\theta - \sin 2\theta) = I_0^2 A \qquad \text{...(11.27)}$$

where θ is the same as defined in equation (11.23). The energy stored in the lumped element is calculated in terms of the current flowing through the element. For an inductor the stored energy is given by

$$U_1 = \frac{1}{2} L (I_0 \sin\theta)^2 = I_0^2 B \qquad \text{...(11.28)}$$

Value of Q_1 is obtained from equations (11.26), (11.27) and (11.28) as

$$Q_1 = B\left(\frac{A+B}{Q_{01}} - \frac{A}{Q_0}\right)^{-1} \qquad \text{...(11.29)}$$

where A and B have been defined in (11.27) and (11.28), respectively.

Coaxial line resonators have been used for these measurements. Half-wave microstrip resonators may also be used, in which case, the lumped element can be series-mounted in the middle of the resonator as shown in Fig. 11.7. Placing of the lumped element in this manner will modify the resonant frequency for those modes of the resonator for which the current has a non-zero value at the location of the element. Fundamental mode of resonance and odd harmonics have this property.

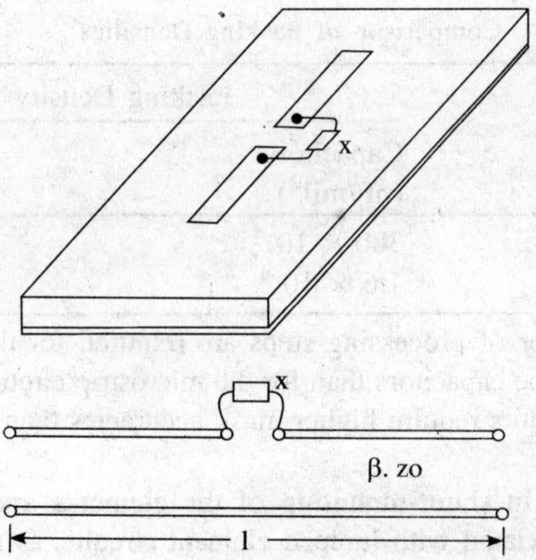

Fig. 11.7

It is also possible to shunt-mount the components in the middle of a transmission line resonator. In this case, only even harmonics of the fundamental will be affected.

11.4 CIRCUITS USING LUMPED ELEMENTS

The design procedures for circuits using lumped elements at microwave frequencies are similar to the corresponding methods for low frequencies. In fact, it is a major advantage of using lumped elements so that the circuit design and optimization techniques developed for low frequency circuits may be extended to microwave frequencies also. Distributed element circuit design may be transformed to the lumped element circuits by using lumped equivalents for transmission line sections.

Several types of circuits using lumped elements have been reported in literature. Some of these are :

1. S-band low pass filter
2. Branch line directional coupler (S-band)
3. C-band impedance transformer
4. Transistor amplifier at S-band
5. Tunnel diode amplifier at 4 GHz
6. X-band varactor tuned Gunn oscillator
7. A degenerate S-band parameters amplifier
8. An X-band doppler radar

The main features of lumped element circuits vis-a-vis the distributed circuits may be summarized as follows:

1. One of the main attractions of lumped elements is their relatively small size. A comparison of packing densities is made in Table 11.1. Because of high packing density, lumped elements are strong contenders for MICs.

TABLE 11.1

Comparison of Packing Densities

Elements	Packing Density	
	Capacitors (pF/mil^2)	Inductors (nH/mil^2)
Lumped	300×10^{-4}	5.5×10^{-4}
Distributed	1.6×10^{-4}	2×10^{-4}

2. More number of processing steps are required for the fabrication of the sandwich type capacitors than for the microstrip circuits. Also, the lumped element circuits require higher mask accuracies than those for microstrip circuits.
3. Difficulties in shunt-mounting of the elements and in ground returns may be alleviated with lumped element circuits, as a ground connection can be placed near. This provides an advantage over microstrip circuits.

4. Lumped element circuit design employs procedures commonly used at lower frequencies. Thus circuit optimization methods used at lower frequencies can be extended to microwave frequencies.
5. Combination of lumped elements with uncapsulated semiconductor devices leads to a significant reduction in price and also improves the bandwidth. This improvement arises not only from the use of devices in chip form but also from the use of lumped resonant structures instead of distributed resonant structure.

The improvement in the bandwidth arises from the fact that the equivalent reactances of the distributed elements are frequency dependent whereas lumped elements have values essentially independent of frequency. For illustration, consider the situation shown in Fig. 11.8. Here, an inductor is used to resonate the capacitance of a semiconductor chip.

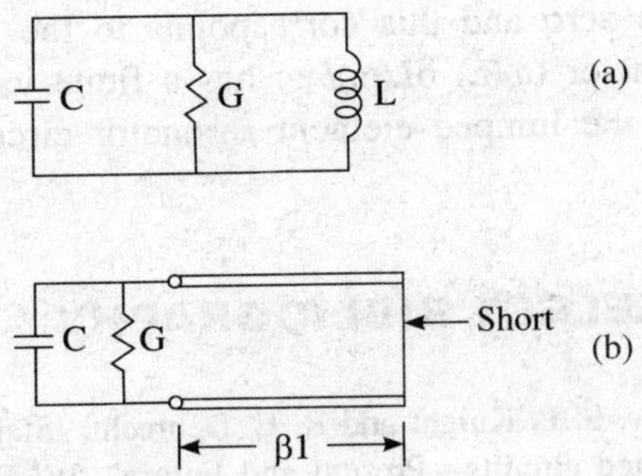

Fig. 11.8

Figure 11.8(a) shows a lumped inductor while Fig. 11.8(b) shows its transmission line equivalent. In general, Q is defined (at resonance) by

$$Q = \frac{\omega}{2G}\left(\frac{\delta B}{\delta \omega}\right) \qquad ...(11.30)$$

where G is the total conductance and $\delta B/\delta\omega$ is the rate of change of susceptance with frequency. Calculations of Q of the lumped circuit (Q_L) and of the corresponding distributed circuit (Q_D) show that

$$Q_D = \frac{Q_L}{2}\left(1 + \frac{2\beta l}{\sin 2\beta l}\right) \qquad ...(11.31)$$

From the above equation we note that the distributed Q is much larger than the lumped Q and it tends to the lumped value when βl is small. This derivation is quite general and demonstrates that for broadband circuits, lumped circuit elements have an advantage over distributed elements.

Another typical example where the lumped elements improve the circuit performance is the case of a varactor tuned semiconductor oscillator.

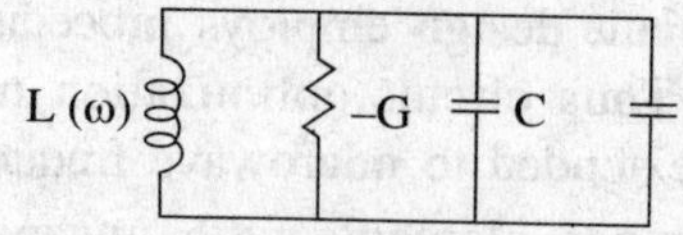

Fig. 11.9

Figure 11.9 shows oscillator circuit with frequency dependent inductance $L(\omega)$. The capacitance change δC is caused by the varactor. In this circuit it is desirable that a given δC should produce a large change in the resonance (and thereby oscillation frequency). It can be shown that the change in frequency $\delta\omega$ for a change in capacitance δC is given by

$$\frac{\delta\omega}{\delta C} = -\frac{1}{2}\frac{\omega/C}{1+(\omega/L)[\delta L(\omega)/\delta\omega]} \qquad ...(11.32)$$

This relation shows that the rate of change of frequency $\delta\omega/\delta C$ has its maximum value when $\delta L(\omega)/\delta\omega$ is zero and this corresponds to the lumped circuit case. If L is distributed, the factor $(\omega/L)\ \delta L(\omega)/\delta\omega$ has a finite value which is usually greater than zero. Thus the lumped element resonator circuits provide a wider tuning range.

SELECT BIBLIOGRAPHY

M. Caulton, B. Hershenov, S. P. Knight and R. E. Debrecht, 'Status of lumped elements in microwave integrated circuits—Present and Future'. *IEEE Trans. on Microwave Theory and Tech.*, MTT-19, July 1971, p. 588-99.

D. A. Daly, S. P. Knight, M. Caulton and R. Ekholdt, 'Lumped elements in microwave integrated circuits', *IEEE Trans. on Microwave Theory Tech.*, MTT-15, Dec. 1967, pp. 713-21.

M. Caulton, S. P. Knight and D. A. Daly, 'Hybrid integrated lumped element microwave amplifiers', *IEEE Trans. Electron Devices,* ED-15, July 1968, pp. 459-66.

M. Caulton, 'The lumped element approach to microwave integrated circuits', *Microwave Jour.*, 13, May 1970, pp. 51-8.

M. Caulton, and H. Sobol, 'Microwave integrated circuit technology—A survey', *IEEE J. Solid State Circuits,* SC-5, Dec. 1970, p. 292-303.

F. E. Terman, *Radio Engineering Handbook,* McGraw-Hill, New York, 1943.

G. D. Alley, 'Interdigital capacitors and their applications', *IEEE Trans.* MTT-18, 1970, pp. 1023-33.

C. S. Aitchison, *et al.*, 'Lumped Circuit elements at microwave frequencies' *IEEE Trans. Microwave Theory Tech.*, MTT-19, Dec. 1971, pp. 928-37.

C. S. Aitchison, 'Lumped components for microwave frequencies', *Philips Tech. Rev.*, 32, No. 9/10/11/12, pp. 305-314 (1970).

PROBLEMS

11.1 A $\lambda/10$ section of an open-circuited transmission line, with parameters given by $r = 0.1$ ohm/cm, $L = 0.1$ nh/cm, g = 0.2 mlli mho/cm and $C = 1$ pF/cm is to be used as a lumped capacitor at 10 GHz. Find the value of capacitance, its equivalent circuit and Q-factor.

11.2 An RF choke is required for designing a bias-tee operating at 10 GHz in a 50-ohm microstrip configuration. This can be achieved by using a lumped inductor as shown in Fig. 11.10 (a). Find the value of the inductance such that VSWR does not exceed 1.01. Determine the layout of this inductor such that the maximum dimension is less than 50 mil. Minimum line width and spacing that can be conveniently etched are 1 mil and 0.6 mil, respectively and the film thickness is 0.3 mil.

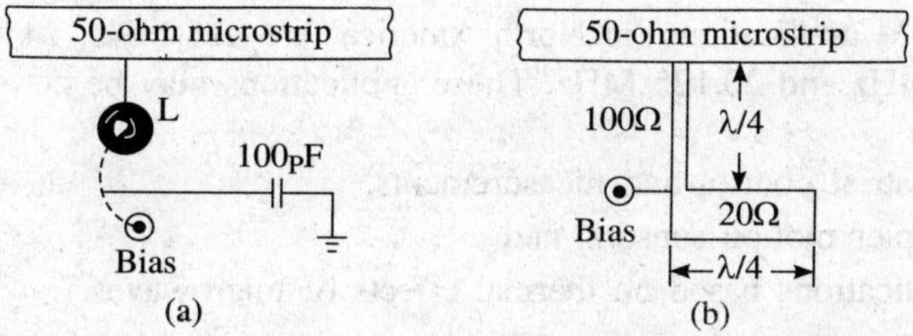

Fig. 11.10

11.3 The RF choke discussed in Prob. 11.2 may also be designed in the form of a microstrip circuit shown in Fig. 11.10 (b). The maximum and minimum impedances of microstrip sections are limited to 100 and 20 ohms, respectively. Compare the VSWR performance of this circuit with the lumped circuit in Problem 11.2 at 8.0 and 12.00 GHz. Ignore the effect of variations of λ_g / λ_0 with frequency.

11.4 Show that for a lumped element series-mounted in the middle of a resonator (Fig. 11.7) value of reactance is given in terms of resonance frequency f as

$$X = 2Z_0 \cot[\pi f/(2f_n)]$$

where f_n is the resonant frequency for the line length l without reactance X being present. Stray reactances at open ends may be ignored.

11.5 Derive the expression for VSWR (Eqn. 11.20) used for measuring high VSWR by double minimum method.

11.6 Derive the expression for Q-factor (Eqn. 11.31) for a resonator using transmission line section as an inductor as shown in Fig. 11.8 (b).

11.7 Consider the equivalent circuit of a varactor tuned oscillator as shown in Fig. 11.9. Show that the change in the frequency of oscillation for an incremental change in the varactor capacitance is given by Eqn. (11.32).

CHAPTER

12

Industrial Applications

Apart from their widespread use in radar and communications, microwaves are finding interesting applications in several other areas. These are known as industrial, scientific and medical (ISM) applications of microwave energy. Frequencies allocated for ISM applications (in North America) are 915 MHz, 2450 MHz, 5800 MHz, 10,525 MHz and 20,125 MHz. These applications may be classified in three groups, namely,

(i) Industrial control and measurements,
(ii) Doppler motion sensors, and
(iii) Applications based on thermal effects of microwaves.

12.1 INDUSTRIAL CONTROL AND MEASUREMENTS

Typical applications that could be grouped under this heading are :

(i) Thickness measurement of metal-sheets in rolling machines.
(ii) Continuous measurement of wire diameters.
(iii) Monitoring and measurement of moisture content.

12.1.1 Measurement techniques

Most of the measurements for industrial applications involve determination of a complex reflection or transmission coefficient at microwave frequencies. Use of microwaves is characterized by the following features :

1. At microwave frequency, skin depth in metals is very small and, therefore, a plane metal surface causes a total reflection of waves. Skin depth* δ is given by

$$\delta = 1/\sqrt{\pi f \mu \sigma} \qquad ...(12.1)$$

2. Wavelengths are small and, therefore, the phase variations are rapid. Consequently, a small change in position (or dimensions) gives rise to a significant phase change that can be detected and measured.

* Skin depth for copper is given by $0.066/\sqrt{f}$ metre.

A typical comparison bridge for measurement of the reflection coefficient is shown in Fig. 12.1. Signal from a microwave source is fed to the *E*-branch of a waveguide hybrid-tee and divides equally into two collinear branches. One of the collinear branches *(A)* is terminated in a tuner and a matched termination while the other branch *(B)* looks into the unknown reflection. This method is very useful in measuring deviations in the reflection coefficient from a standard value. The tuner connected to *A* is adjusted such that the reflected power in two branches are equal and in phase for the standard reflection at *XX*. As a result, there is no signal in the *H*-arm. Any change in the reflection coefficient at the plane *XX* causes a signal to propagate in the *H*-arm. This is detected, amplified and displayed on a calibrated meter.

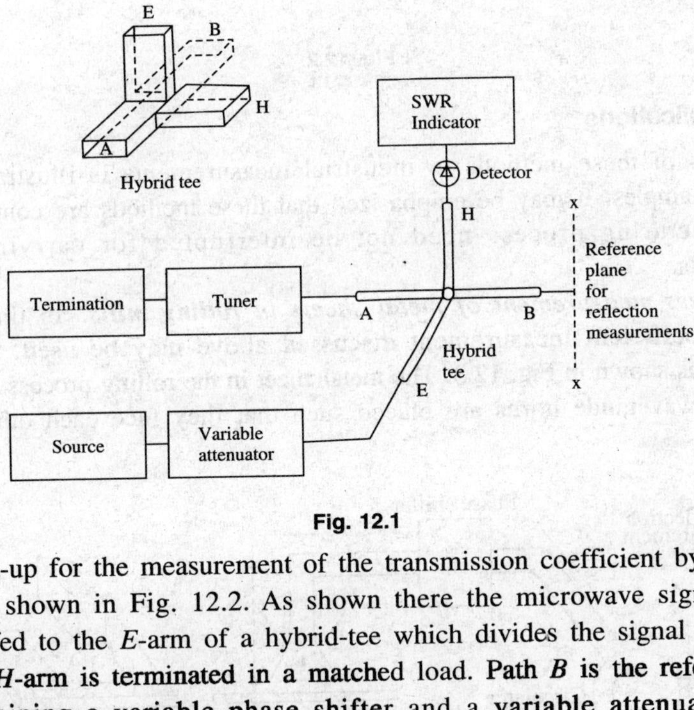

Fig. 12.1

A set-up for the measurement of the transmission coefficient by comparison method is shown in Fig. 12.2. As shown there the microwave signal from the source is fed to the *E*-arm of a hybrid-tee which divides the signal in two paths *A* and *B*. *H*-arm is terminated in a matched load. Path *B* is the reference signal path containing a variable phase shifter and a variable attenuator. Signals transmitted through paths *A* and *B* are fed to two collinear arms of the second hybrid-tee at *A′* and *B′*, respectively. The attenuator and the phase shifter in path *BB′* are adjusted such that for the reference situation in path *AA′*, the two signals reaching *A′* and *B′* are equal in magnitude and are in phase. Thus there is no power coupled to *H*-arm and the detector output indicates a null. If there is any change in transmission coefficient of the circuit in path *AA′*, the difference signal is coupled to the *H*-arm of the second hybrid and can be displayed on a suitably calibrated meter.

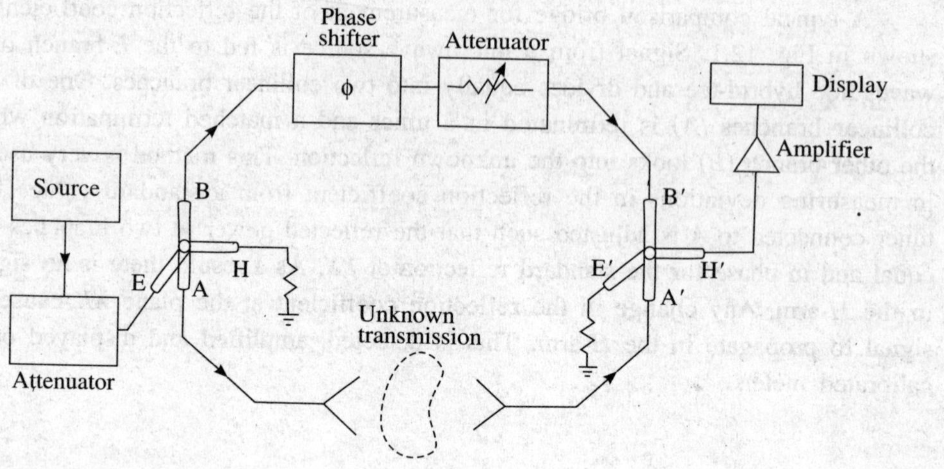

Fig. 12.2

12.1.2 Applications

Applications of these methods to industrial measurements is illustrated by the following examples. It may be emphasized that these methods are contactless and the manufacturing process need not be interrupted for carrying out the measurements.

(a) ***Thickness measurement of metal sheets in rolling mills:*** In this case the reflection coefficient measurement discussed above may be used. A possible arrangement is shown in Fig. 12.3. The metal sheet in the rolling process is indicated by *M*. Two waveguide horns are placed such that they face each other and the

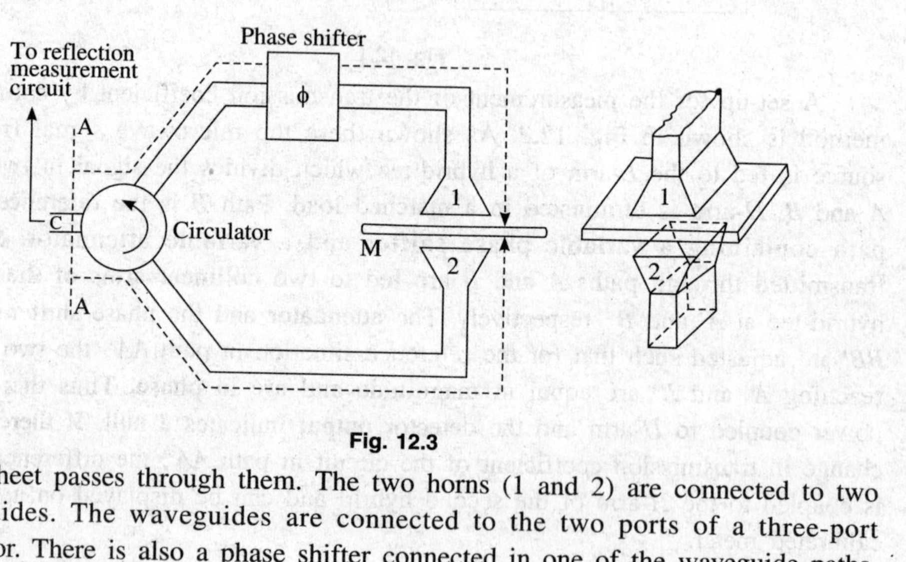

Fig. 12.3

metal sheet passes through them. The two horns (1 and 2) are connected to two waveguides. The waveguides are connected to the two ports of a three-port circulator. There is also a phase shifter connected in one of the waveguide paths.

The third port of the circulator is connected to a coupling element *C*. Length of the waveguide along with circulator, phase shifter and horns, shown dotted in the Fig. 12.3 constitutes a resonator cavity. Its resonance frequency can be varied by adjusting the phase shifter and is set equal to that of the signal source. This sets the reflection coefficient at plane *AA* to a small value. Any change in the thickness of the metal plate *M*, changes the length of the cavity and thus, the resonance frequency. This can be easily detected as a change in the reflection coefficient at plane *AA*. Thus a suitably calibrated meter can be used to monitor thickness. It may be noted that a small displacement in the position of the metal plate does not change the overall length of the resonator and hence does not affect the thickness measurements.

(b) Measurement of wire diameter in drawing operations: Transmission coefficient measurement discussed earlier may be used for this purpose also. A possible arrangement is shown in Fig. 12.4. The wire *W* passes through the space between two closely spaced waveguide horns 1 and 2. Transmission from the plane *A* to the plane *B* is measured by the set-up shown in Fig. 12.2. Since the power radiated by the horn 1 gets diffracted by the wire and reaches the horn 2, a change in wire diameter alters the amplitude and the phase of the signal reaching the horn 2. These changes in the amplitude and the phase can be easily detected by the set-up described earlier.

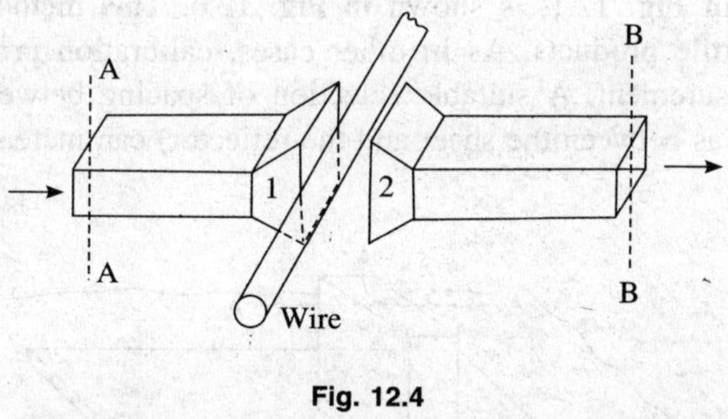

Fig. 12.4

(c) Thickness measurement of dielectric sheets : This measurement can be carried out by a suitably designed transmission set-up very similar to that used for the measurement of wire diameter. As shown in Fig. 12.5, the dielectric sheet passes through the space between the two horns. If the thickness of the sheet varies, transmitted power is changed and can be detected. The indicating meter can thus be calibrated in terms of the sheet thickness.

(d) Monitoring of moisture content in paper and textile industry: For the measurement of moisture content, various types of microwave bridges have been used. The key feature of these methods is to evaluate the attenuation and phase

shift of the electromagnetic wave passing through the material or reflected from it. The techniques differ slightly from one another according to the structure and shape of the test object; e.g. set-up for tests on sheet material (paper, textile, laminated wood, etc.) differs from that designed for tests on granulated or porous material (sand, clay, flour, corn or bricks) or liquids.

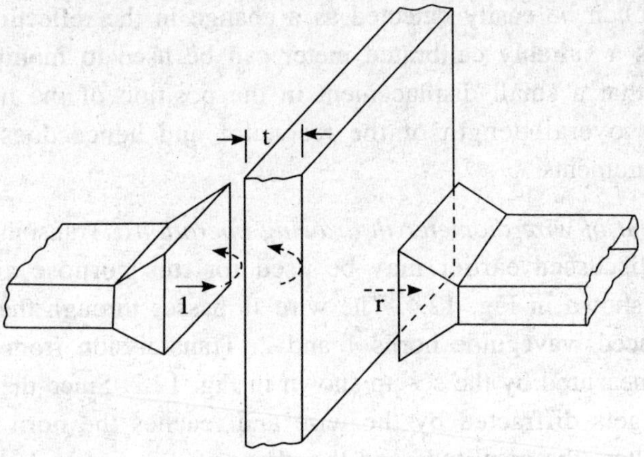

Fig. 12.5

A reflection measurement set-up for sheet materials, to be used along with the arrangement of Fig. 12.1, is shown in Fig. 12.6. This method is well suited for paper and textile products. As in other cases, calibration procedure is a key step for this measurement. A suitable selection of spacing between the horn and the sheet (as well as between the sheet and the reflector) can increase the sensitivity of the method.

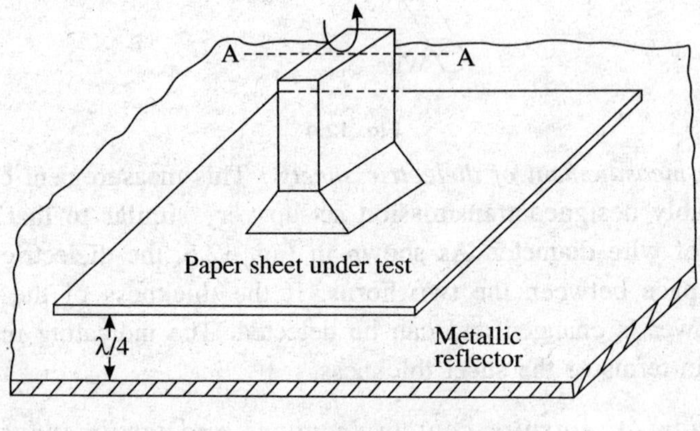

Fig. 12.6

It may be noted that the above method indicates the average moisture content of the material present within the zone of measurement while conventional methods determine the moisture content on small discrete samples.

(*e*) *Measurement of moisture content in liquids* : The arrangement shown in Fig. 12.6 is not suitable for liquids. Several alternatives are possible. In one of these the liquid is passed through a thin glass tubing and the moisture content is monitored by the change it produces in the effective permittivity of the liquid. This can be done by making the tube (through which the liquid is passing) a part of a microwave cavity resonator and measuring the resonant frequency of the cavity. The cavity design is chosen such that the microwave E field is maximum along the axis of the tube. Any change in permittivity of the fluid causes a change in the resonance frequency of the resonator. This change can, however, be detected by using the set-up shown in Fig. 12.7. Details of the cavity are shown in Fig. 12.8.

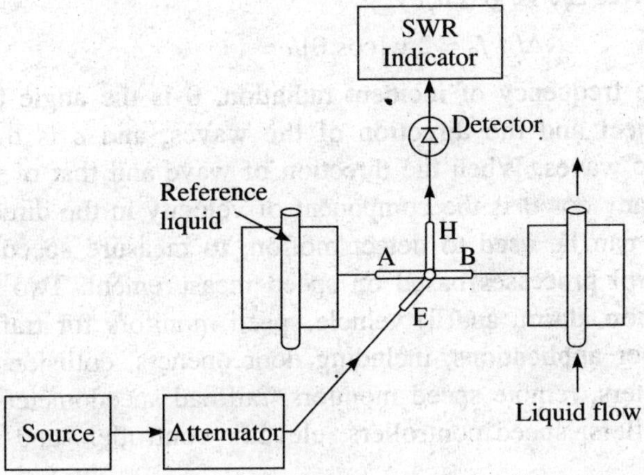

Fig. 12.7

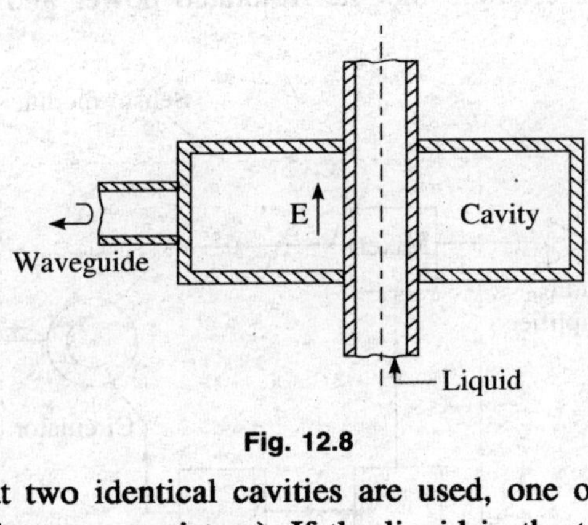

Fig. 12.8

It may be noted that two identical cavities are used, one of which contains the reference liquid (without any moisture). If the liquid in the test cavity is also free from moisture, the resonance frequency of two cavities will coincide and the detector meter will show a null. This is because the reflected powers from the two ends of the collinear branches in this case become equal. When the resonance frequency of a cavity changes, its reflection coefficient changes too and the bridge will thus no longer be balanced. Detector output can be calibrated in terms of moisture content. Examples given above have been selected to illustrate the basic principles of the applications of microwaves in industrial control and measurements. A variety of similar techniques can be designed to tackle the various measurement and control problems.

12.2 DOPPLER MOTION SENSORS

Doppler motion sensors use basically the Doppler effect, viz., change in the frequency of waves when they are reflected by a moving object. When an object is moving with a velocity v towards a source of radiation, the shift in the frequency of reflected wave, Δf, is given by

$$\Delta f / f_0 = 2\, v\, [\cos \theta]/c \quad ...(12.2)$$

where f_0 is the frequency of incident radiation, θ is the angle between trajectory of moving object and the direction of the waves, and c is the velocity of the electromagnetic waves. When the direction of wave and that of the moving object are not collinear,$v \cos \theta$ is the component of velocity in the direction of the wave. This principle can be used to detect motion, to measure speed and to carry out industrial control processes based on speed measurement. Two main applications are : (i) intrusion alarm, and(ii) vehicle speed monitors for traffic control. There are several other applications, including door openers, collision warning systems, auto speedometers, remote speed monitors, railroad speedometers, traffic counters, process controllers, speed controllers, elevator controllers and runway monitors, etc.

Basic block diagram of a microwave Doppler sensor is shown in Fig. 12.9. Microwave circuit portion is shown in dotted block and could be fabricated in hybrid MIC form as a single module. Radiated power and local oscillator power

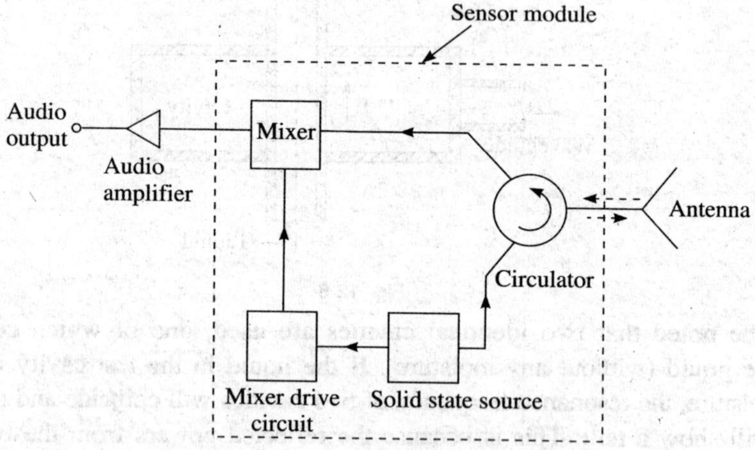

Fig. 12.9

for the mixer is obtained from the same source. Thus any drift in the source frequency f_0 does not affect the measurement of Doppler shift Δf. Power required for any specific application can be determined from the radar equation. Usually power levels of 10 to 25 mW are sufficient for most of the applications and can be obtained from transferred electron devices discussed in Chapter 4.

Out of the various frequencies in ISM band, 10.525 GHz is perhaps best suited for Doppler sensor application. At lower frequencies the antenna size becomes large and at higher frequency (22.125 GHz) the components and the sources are more difficult to obtain. Also, for 22 GHz band, human-motion Doppler frequencies lie near the power line frequencies.

12.3 APPLICATIONS BASED ON MICROWAVE HEATING

Processing and cooking of food products is a major application of dielectric heating at microwave frequencies. Apart from food processing, there are other applications of microwave heating. These include : (i) medical applications (diathermy, cancer treatment, etc.), (ii) sealing of plastics, (processing of dielectrics, etc.), (iii) destruction of dry rot, fungus and wood worms in timber, (iv) curing and breaking of concrete, and many others. The following discussion is mainly concerned with food processing.

Of the ISM frequencies, presently available for industrial applications, only 915 and 2450 MHz are of importance for microwave heating applications, especially for processing of food products. Power generation becomes more costly at higher microwave frequencies and very little background information in terms of dielectric data, or heating experiment data exist at higher frequencies.

12.3.1 Principles of microwave heating

In conventional methods of heating, heat is transferred to the surface of the material to be heated by conduction, convection, and/or radiation, and into the interior mainly by thermal conduction. In contrast, in microwave heating (or the dielectric heating at any other frequency) heat is generated directly inside the material. Therefore, it is possible to obtain a much faster temperature rise than in conventional heating. However, heat conduction still plays an important role when heating thick samples as it equalizes temperatures when the heat generation is uneven.

When a material is exposed to microwave radiations, fields penetrate to a depth given by skin depth. Dielectric losses which cause heat generation are proportional to $\in_r''$, the imaginary port of the dielectric constant. The power dissipated in a certain volume V of the material is given by

$$P_d = \frac{1}{2} \in \omega_0 \int_V \in_r'' |E|^2 \, dV = 27.8\times10^{-12} f \int_V \in_r'' |E|^2 \, dV \qquad ...(12.3)$$

where f = frequency in Hertz,

E = amplitude of electric field in volts/m,

$\varepsilon_r = (\varepsilon_r' - \varepsilon_r'')$, relative permittivity of the material, and

P_d = power dissipated in watts.

Power dissipated in the material determines the temperature rise at any point according to the relation

$$\frac{\partial T}{\partial t} = 0.239 \frac{P_d}{c\rho} \text{ (°C/sec)} \qquad ...(12.4)$$

where c is the specific heat of the material in calories per gram and ρ the density of the material in grams per cubic centimetre.

From (12.3) we note that the dissipated power is proportional to the frequency, so the higher the frequency, more is the power dissipated. An additional advantage of microwave heating results from the behaviour of ε_r'' as a function of frequency. Most of the food products contain free water and the dipolar dispersion in water occurs at microwave frequencies. Value of ε_r'' increases because of the dipolar dispersion. For free water at 25°C, ε_r'' is 0.36 at 10 MHz, but increases to 12 at 3 GHz. This shows that the increase of dissipated power at microwave frequencies can be substantial.

12.3.2 Applicators for microwave heating

A device in which microwave energy interacts with the material being heated is known as a *microwave applicator.* A good applicator should assure conversion of the delivered microwave energy into thermal energy in the most effective manner. Various possible configurations for applicators may be divided into three groups : (a) multimode cavity type, (b) open slow-wave-structure type, and (c) waveguide type.

MULTIMODE CAVITY TYPE APPLICATORS

In multimode cavity, available power is distributed in as many modes as possible within a closed volume. Access to this type of system is obtained either through ports which are designed to act as wavetraps, or through special type of doors. This system is most versatile for all but thin materials in sheet form.

In multimode cavity design, two requirements predominate. First, the source coupling to the cavity has to remain constant over a wide range of load, the dielectric properties of which change with temperature. Secondly, energy conversion must occur uniformly in the material. Wall losses are reduced by using minimum metal surface area for a given volume. The stability and uniformity of heating depend primarily on a sufficiently large number of modes present in the cavity under all loading conditions.

The resonant frequencies of transverse electric (TE) and transverse magnetic (TM) modes in a cavity of dimensions $(a \times b \times d)$ is given as follows (*see* also Eqn. 2.92).

$$\left(\frac{2\pi f}{c}\right)^2 = \left(\frac{\pi m}{a}\right)^2 + \left(\frac{\pi n}{b}\right)^2 + \left(\frac{\pi p}{d}\right)^2 \quad ...(12.5)$$

where *m, n, p* are integers corresponding to TE_{mnp} or TM_{mnp} modes and f is the resonant frequency. Maximum number of modes at a given frequency of excitation occurs for a cubic cavity where $a = b = d$. Then

$$m^2 + n^2 + p^2 = 4(d/\lambda)^2 = q \quad ...(12.6)$$

where q is a positive integer. When two modes (corresponding to different combinations of *m, n* and *p*) have the same resonant frequency, they are said to be degenerate. Since both TE and TM modes are present, the total number of modes is obtained by doubling the distinct permutations of each bracketed set *(m, n, p)*. Distinct permutations of each *(m, n, p)* set correspond to first type degeneracies. New values of *(m', n', p')* are found in terms of unprimed quantities by intercharges given by any of the following three equations and correspond to the second type degeneracies

$$(m')^2 = n^2 + p^2\ ;\ (n')^2 + (p')^2 = m^2 \quad ...(12.7a)$$

$$(n')^2 = p^2 + m^2\ ;\ (p')^2 + (m')^2 = n^2 \quad ...(12.7b)$$

$$(p')^2 = m^2 + n^2\ ;\ (m')^2 + (n')^2 = p^2 \quad ...(12.7c)$$

The third type degeneracies arise when *(m', n', p')* are related to the original solution *(m, n, p)* through

$$m^2 + n^2 + p^2 = (m')^2 + (n')^2 + (p')^2 \quad ...(12.8)$$

The first appreciable mode density occurs in the range $49 \le q \le 54$. As an example, when $a = b = d = 43.2$ cm (17 inch cube), then for $2.4 \le f \le 2.5$ GHz, $48 \le q \le 52$ and the total number of modes is 68.

From the above discussion we observe that it is possible to determine optimum dimensions for cubic and rectangular cavities for a given source bandwidth.

Multimode cavity ovens may be modified if heating of continuously flowing materials is required. Two examples are shown in Fig. 12.10. In the first, the material is on a conveyer belt, while in the second meterial flows inside a circular dielectric tube which passes through the cavity. Both of these arrangements require chokes near the walls to ensure that microwave energy does not leak out.

SLOW-WAVE-STRUCTURE TYPE APPLICATORS

The most common example of this type of applicators is meander waveguide structure shown in Fig. 12.11. In this design, a rectangular waveguide carrying the microwave power is folded in the 'meander' shape. Holes in the narrow wall of the waveguide allow microwave power to leak out. Material to be heated is placed on a conveyer belt which moves close to the surface of the meandered waveguide. Density of microwave power available at the surface of the conveyer belt can be controlled by the size and the number of holes in the waveguide wall. Also, a temperature gradient in the direction of flow can be designed, if desired.

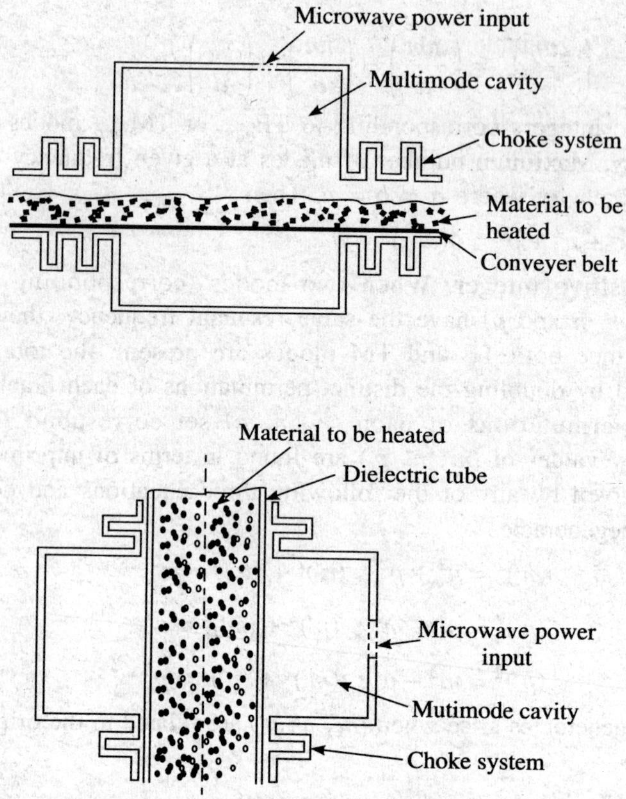

Fig. 12.10

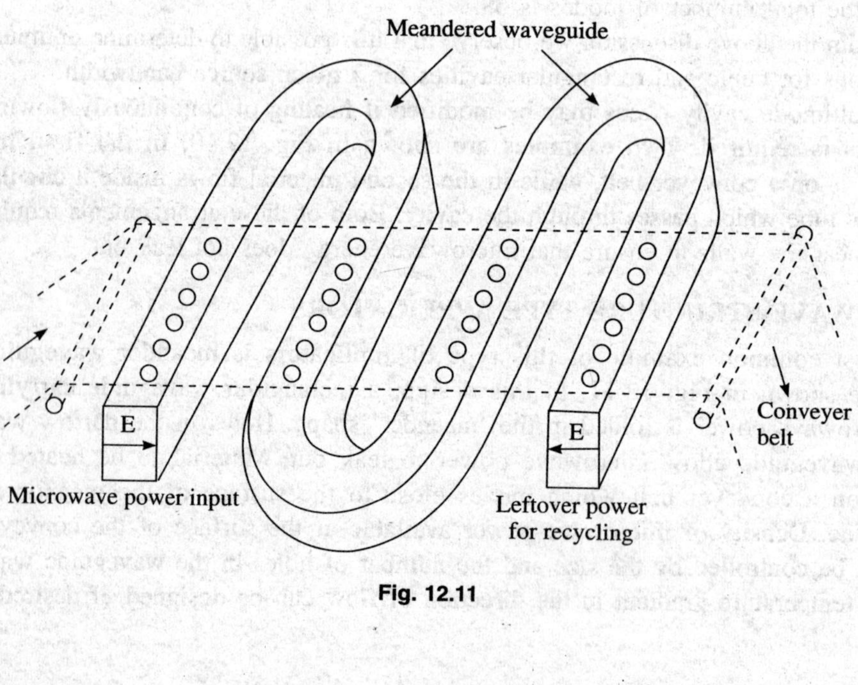

Fig. 12.11

The meandered waveguide structure discussed above becomes a slow-wave-structure for microwaves, since the velocity of waves in the direction of movement of conveyer belt is less than the velocity of light. Several other designs for slow-wave-structures (like a corrugated metal surface) could also be used in this type of applicators. It may be recalled that the slow-wave-structures are used in travelling wave tubes also (*see* Sec. 3.5), and have been studied extensively in that context.

WAVEGUIDE TYPE APPLICATORS

Circular and rectangular waveguides as well as coaxial lines may be used as microwave applicators. Waveguide is filled with the material to be heated and the microwave energy is passed through the filled waveguide. Material to be heated can flow continuously through the waveguide. An example of this type of applicators is shown in Fig. 12.12. It uses a dielectric layered circular waveguide. Presence of the dielectric layer helps in reducing the non-uniformity of the dissipated power across the cross-section of the waveguide.

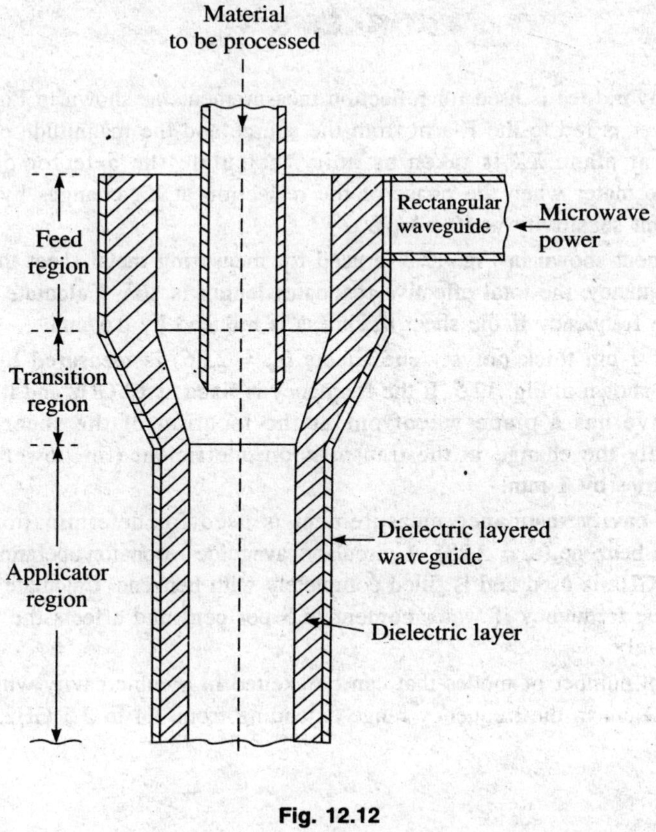

Fig. 12.12

SELECT BIBLIOGRAPHY

E. C. Okress (ed.), *Microwave Power Engineering,* Vol. 2, Academic Press, 1968.

S. Stuchly, A. Kraszowski and M. Rzepecka, 'Microwaves for continuous control of industrial process', *Microwave Jour.,* 12, No. 9 (Aug. 1969), p. 51.

'Special issue on the industrial, scientific and medical applications of microwaves', *Proc. IEEE,* 62, No. 1. Jan. 1974.

R. H. Merriam and J. W. Rush, 'Microwave doppler sensors', *Microwave Jour.,* 17, No. 7 (July 1974), pp. 27-30.

K. C. Gupta, 'Cylindrical waveguide applicators with uniform power flow distributions', *Journal of Microwave Power,* 11, No. 4 (Dec. 1976), pp. 321-31.

M. A. Rzepecka, S. S. Stuchly and M. A. K. Hamid, 'Applicators for microwave treatment of granular materials in agricultural technology', *IMPI Tech. Rept.* Series-TR-72-1, (1972).

K. C. Gupta and I. J. Bahl, 'RF applicators find jobs—on farms, in factories', *Microwaves,* 17, (June 1978), pp. 52-64.

PROBLEMS

12.1 A matched hybrid-tee is used for reflection measurements as shown in Fig. 12.1. If a 10 mW power is fed to the E-arm from the source and the magnitude of reflection coefficient at plane *XX* is taken as unity, calculate the detector output on a microampere meter when the phase of the reflection at *XX* changes by 10° (short circuit current sensitivity is 10 μA/μW).

12.2 An arrangement shown in Fig. 12.3 is used for measuring metal sheet thickness. At 10 GHz frequency, the total effective resonator length is 10λ. Calculate the change in resonance frequency if the sheet thickness is reduced by 0.5 mm.

12.3 Thickness of 1 cm thick polystyrene sheets (ε_r = 2.56) is measured by using the arrangement shown in Fig. 12.5. If the frequency is fixed at 10 GHz and it is assumed that the wave has a plane wavefront at the location of the sheet, calculate approximately the change in the transmission coefficient (for power) when the thickness varies by 1 mm.

12.4 Microwave cavity resonance measurement is used for determination of water (ε_r = 80) in benzene (ε_r = 2.28). A circular waveguide resonator operating in TM_{010} mode at 10 GHz is used and is filled completely with benzene. Calculate the shift in the resonance frequency if water content is 5 per cent and affects the value of ε_r proportionately.

12.5 Find the total number of modes that can be excited in a cubic cavity with each side equal to 43.2 cm in the frequency range extending from 2.4 to 2.5 GHz.

Appendix

Physical Constants

Velocity of electromagnetic waves in free space	$c = 2.998 \times 10^{8}$ m/sec
Charge of the electron	$e = 1.602 \times 10^{-19}$ coulomb
Mass of the electron	$m = 9.108 \times 10^{-31}$ kg
Charge-mass ratio of the electron	$e/m = 1.76 \times 10^{11}$ coulomb/kg
Permittivity of free space	$\varepsilon_0 = 8.854 \times 10^{-12}$ farad/m
	$= \frac{1}{36\pi} 10^{-9}$ farad/m
Permeability of free space	$\mu_0 = 4\pi \times 10^{-7}$ henry/m
Wave impedance of free space	$\eta = 376.7 = 120\pi$ ohms
Boltzmann's constant	$k = 1.380 \times 10^{-23}$ joule/°K
Planck's constant	$h = 6.547 \times 10^{-34}$ joule-sec
Electron plasma frequency (N = number of electrons/m^3)	$f_p = \frac{e}{2\pi}\left(\frac{N}{m\varepsilon_0}\right)^{1/2}$
	$= 8.97\ N^{1/2}$ Hz
Electron cyclotron frequency	$f_c = eB/(2\pi m)$
(B in weber/m^2)	$= 28{,}000B$ MHz/sec
Conductivity of copper	5.8×10^{7} mho/m

Index